Shale Gas:
The Promise and the Peril

by Vikram Rao

D1456181

RTI

RTI
NATIONAL

Library of Congress Control Number: 2012941607

ISBN 978-1-934831-10-6
(refers to printed version)

RTI Press publication No. BK-0009-1206
doi:10.3768/rtipress.2012.bk.0009.1206
www.rti.org/rtipress

Cover design: Tayo Jolaoso
Pen and ink illustrations: Gordon Allen

The RTI Press mission is to disseminate information about RTI research, analytic tools, and technical expertise to a national and international audience. RTI Press publications are peer-reviewed by at least two independent substantive experts and one or more Press editors.

RTI International is an independent, nonprofit research organization dedicated to improving the human condition by turning knowledge into practice. RTI offers innovative research and technical services to governments and businesses worldwide in the areas of health and pharmaceuticals, education and training, surveys and statistics, advanced technology, international development, economic and social policy, energy and the environment, and laboratory testing and chemistry services.

This publication is part of the RTI Press Book series.

RTI International

3040 East Cornwallis Road, PO Box 12194, Research Triangle Park, NC 27709-2194 USA

rtipress@rti.org

www.rti.org

Dedication

To my grandmother, Srimati Manorama Bai, a pioneer educator in South India. She would have been pleased.

Contents

Glossary

Aquifers. Underground bodies of water. Freshwater aquifers, through drilled water wells, are the primary source of water for human consumption other than surface water.

Breaker. A chemical used to eliminate the effects of the cross-linker, thus allowing the viscous fluid to be retrieved from the reservoir more easily.

Brackish. Water with soluble salts such as chlorides in greater concentration than fresh water (500 ppm) but less than sea water (35,000 ppm).

BTEX. Benzene, toluene, ethylbenzene, and xylene. These are volatile organic compounds whose chemical structure causes them to be designated as aromatics. This term was originally coined because these compounds tend to have an aroma, albeit not a pleasant one.

CERA. Cambridge Energy Research Associates. One of the most reputable energy consulting houses, best known for their annual conference in the spring.

CNG. Compressed natural gas. Gas in this form is stored at about 3,000 psi to 3,500 psi and used for a variety of purposes.

Crack spread. The price difference between the original fluid and the final cracked fluid. After processing costs, this is reflective of the profit in the operation.

Cracking. A refining process for converting larger molecules to smaller, more useful ones. This can be done thermally and/or by using specialized catalysts.

Cross-linker. A chemical used in some fracturing operations to make the fracturing fluid more viscous. Doing so increases the efficacy of producing fractures in the rock.

Crude oil. A mixture of naturally occurring hydrocarbons which can be processed in refineries to yield useful fluids such as gasoline, diesel, and jet fuel. The bulk of crude oil has the formula C_nH_{2n+2}, where n is generally over 20.

CTL. Coal to liquids. This term describes the process for converting coal into liquid hydrocarbons. The original application of the Fischer-Tropsch process was CTL.

Diesel. A crude oil distillate that can be compression-ignited, as opposed to spark ignition, required for gasoline.

DME. Dimethyl ether. A simple ether derived from methanol. It can be blended with diesel to at least 20 percent by volume without engine modifications. It has a higher cetane rating than diesel and produces zero particulates when combusted.

Drop-in fuel. This is a class of fuels which may be blended with, or "dropped into," conventional automotive fuels. In the pure definition, this ought to be possible in any proportion.

Dry gas. Natural gas that is substantially free of natural gas liquids.

FFV. Flex fuel vehicle. A vehicle with an engine that can operate on any mix of gasoline and alcohol.

Fischer-Tropsch. Abbreviated F-T or FT, this is one of the most common gas-to-liquid processes.

Flocculent. A fluffy material which, in the context of water treatment, captures undesirable species and is then discarded.

Flowback water. Water used in the fracturing operation that is circulated back to the surface after the operation is complete. It will often contain some proportion of water from the formation. In shale oil and gas operations, the flowback water is decidedly more salty than what went in. This is distinct from *produced water*, which is *formation water* that flows after the hydrocarbon is mostly extracted. They do overlap, in that the flowback water will have a component of formation water.

Formation water. The natural water layer in a natural gas or coal reservoir.

Fossil fuel. Fuel consisting of the remains of organisms preserved in rocks in the earth's crust with high carbon and hydrogen content.

Fracturing. Defined as an operation in which high-pressure fluid, usually water-based, is injected into reservoir rock in order to fracture it to induce artificial permeability.

F-T. See Fischer-Tropsch.

Fugitive gas emissions. Methane released inadvertently into the atmosphere or into a body of water.

Gasoline. A crude oil distillate that powers internal combustion engines; can also be synthesized from natural gas.

GTL. Gas to liquids. The process of converting natural gas to useful liquids, usually used in connection with producing transport fuel.

Henry Hub price. The price in Erath, Louisiana, for natural gas, used for trading on the New York Mercantile Exchange. It is strongly correlated with the price of gas in all parts of the US.

Interval (well). A well interval is a portion along the length of the well. The designation is used to identify zones of different character, such as productivity.

IHS. A large consulting house of which CERA (Cambridge Energy Research Associates) became a part in 2004.

Kerogen. A mixture of long-chain organic molecules that is a precursor of oil and gas.

LNG. Liquefied natural gas. Natural gas chilled to the fluid state and kept chilled at about -260°F. Its volume is 600 times less than free gas. Transport of natural gas across the ocean is done in LNG tankers.

LPG. Liquefied petroleum gas. A mixture of molecules larger than ethane and methane. Most LPG is some blend of propane and butane.

Marcellus. The name of the formation of sedimentary rock in the eastern United States containing important deposits of shale gas.

Methane. A colorless, odorless, flammable gas with the formula CH_4. It is the principal constituent of natural gas.

MMBTU. Million British thermal units. This is the most common energy unit to describe the energy content of fossil fuels. A thousand cubic feet of natural gas nominally contains 1 MMBTU.

MTG. Methanol to gasoline. A specific GTL process which converts syngas to methanol and then to gasoline. The syngas may be from any carbonaceous source but is usually natural gas.

NGL. Natural gas liquids. Molecules larger than methane found in association with methane in natural gas. In ascending order of size they are ethane, propane, and butane, plus larger molecules beyond.

Oil shale. A shale rock infused with immature oil and kerogen. Can be thermally processed to yield useful hydrocarbons. This is not to be confused with *shale oil*, which is separately defined below.

Pad drilling. A relatively new type of drilling wherein anywhere from 5 to 40 wells may be drilled from a single location, or "pad." The advantages include less overall road construction and traffic. It also facilitates water treatment operations due to economies of scale.

Permeability. The property of a rock that defines ease of mobility of fluid through it.

Petroleum. The technically exact definition is a mixture of naturally occurring hydrocarbons that includes oil and gas. So, for example, the petroleum industry encompasses both fluids. However, in the parlance *petroleum* is more commonly used synonymously with oil, which is liquid at room temperature.

Play (gas play). A region being mined for petroleum, or the activities associated with its development.

Produced water. Formation water that usually flows out after much of the hydrocarbon has been produced. It is distinct from flowback water, but a portion of flowback water is formation water. Industry nomenclature is not clear on this point, and sometimes flowback water is referred to as "produced water."

Proppant. A hard ceramic material, usually sand, which is placed in the fractures created in the reservoir rock. The purpose is to "prop" the fractures open to allow gas or oil to flow. Absent the proppant, the weight of the thousands of feet of rock above would close the fractures.

Refracturing. Process of fracturing a reservoir a second time in an existing well bore. This is usually conducted a few years after producing gas or oil from the original fractures.

Reserves estimate. An estimate of the hydrocarbon resource which is economically recoverable using current technology. In most cases this estimate will increase as the resource is developed, especially in new plays such as shale gas.

Resource estimate. An estimate of the hydrocarbon likely to be in place, whether economically recoverable or not.

RPSEA. Research Partnership to Secure Energy for America. This organization was formed with congressional line-item funding and includes most players in the oil and gas business. Initial targets were ultra-deepwater and unconventional resources.

Saline aquifers. Aquifers with water too salty for human consumption, generally defined as having greater than 500 parts per million chlorides. Saline aquifers are deeper than freshwater bodies. The water from this source is sometimes referred to as brackish water.

Scale. A crusty deposit formed by compounds of calcium, magnesium, and barium, to name the principal species. Collectively these are classified as divalent ions. In common household usage their removal is known as water softening.

Shale. Rock formed by the deposition of sediment, usually transported by water, and comprising primarily clay and silt. This type of rock is usually deposited in characteristic layer patterns.

Shale gas. Natural gas found in shale bodies.

Shale oil. A mature form of oil in shale bodies, not to be confused with *oil shale*, which is shale that contains kerogen.

Shale play. Shale formations containing significant accumulations of natural gas.

Slickwater fracturing. This term is used when the fracturing fluid used has substantially no gelling agent, such as sugars. Consequently neither cross-linkers nor breakers will be in use. Much of shale gas drilling uses this technique.

Source rock. A shale body with oil or gas in fully mature form, which usually is produced only after it has migrated to a porous body such as sandstone or calcium carbonate.

Spur line. Short-distance pipeline connecting the producing rig site with a major gas export line.

Syngas. Short for synthesis gas. A mixture of carbon monoxide and hydrogen produced from any carbonaceous source, often natural gas. Syngas is the basic building block for a host of chemicals, including methanol and ethers.

Thermal cracking. A refining process involving heat to break down molecules to smaller sizes.

Thermal maturity. Relates to the process of heat and time breaking down organic molecules to first form kerogen and then oil and gas. The stage at which any particular reservoir is in this process is described as the thermal maturity of the reservoir.

Tight gas. Natural gas found in substantially impermeable rock. Shale gas falls in this class, as does gas in sandstones and carbonates. The key distinguishing feature is the difficulty of movement of fluids through the rock. Fracturing the rock induces artificial permeability.

Total organic content. Usually abbreviated to TOC. A measure of the economic potential of hydrocarbon-bearing shale bodies.

Wet gas. Natural gas with enough associated natural gas liquids to materially improve the economics of recovery.

Introduction

Shale gas is a newly discovered resource that could make the US self-sufficient in natural gas for over a hundred years. It has already enabled the US to overtake Russia as the largest producer in the world. It is uniquely enabled by a technology known as hydraulic fracturing, a technique under heavy scrutiny for the risk of environmental damage. Some seek to ban the method, and entire states and one country have issued moratoria.

The primary purpose of this book is to shed light on every issue that I consider germane to shale gas and the enabling technology, fracturing. My intent is for supporters and opponents to learn something they did not know about the issues before reading the book. Perhaps it will cause some to rethink their positions. I know that in researching the issues I became aware of some facets that surprised me. One was the positive impact of shale gas on national security. Another was an understanding of why inadvertent releases of methane occur during gas production.

Others will find ammunition to support their held beliefs. That is fine as well. At least the debate will be joined with factual support. A more informed debate is all one could hope for. This is a critically important issue for the nation, and the world by extension, and we must get this one right.

Responsible production of shale gas offers the promise of low-cost energy for a long time, and not just in the US. At a Department of Energy conference in February 2012, Bill Gates emphasized the connection between cheap energy and improvement in the human condition: "If you want to improve the livelihoods of the world's poorest 1 billion, their access to cheap energy determines if they can afford fertilizer, transport and lighting—the things we take for granted as part of our lives and our dignity. Without advances in energy, they will remain stuck where they are" (as cited in Johnson, 2012).

In this context, telling is the fact that India imports natural gas at a price that in April 2012 was nearly *six times* the price in the comparatively affluent US. Domestic shale gas would make a dent in Indian costly imports. By learning to responsibly produce shale gas, the US is in a position to materially improve the human condition. This book takes a modest crack at discussing all the elements required to achieve this result.

The book is written for general consumption and so is necessarily short on technical depth. But underlying scientific rigor is certainly the intent. Calculations and chemistry details are relegated to shaded boxes, which can be skipped without losing the main thread. Also, throughout the book the term "gas" is used solely to mean natural gas. When gasoline is being discussed, the word is spelled out. The common American term gas is not used in that context.

The book approaches the topic from the standpoint of the economic value, the environmental hazards and potential remedies for them, and the political dimension. Balance is sought in the discussions, but ultimately balance is in the eye of the reader. If I have a bias it is toward seeing problems as opportunities to devise solutions and create economic value. Other than that, I have the bias of most people, of ensuring access to affordable energy while minimizing environmental risk.

The discussion is primarily US-centric because the intended audience is largely domestic. However, the principles apply elsewhere, and where relevant those implications are discussed, such as in chapter 19, "Kicking Shale into the Eyes of the Russian Bear," which also discusses the use of gas supply as a political tool. Much of the world is watching to see whether we can exploit this economic windfall with minimal negative impact on our environment and way of life.

Much of the book appears to have a focus on the present. In part this is deliberate. The realizations regarding the size of the resource, the impact on the economy, and the hazards associated with potentially careless production are relatively recent. Public anxiety is at a higher level than it was after the *Deepwater Horizon* oil spill two years ago. Decisions made today will affect our economic and environmental security and well-being for decades. So a focus on the present is valid. However, the book also contains discussions about the future—in particular, to the predictions of the price of natural gas and oil in the coming years, which have significant bearing on decisions in the coming years. The future is also discussed in the context of alternative fuels, which necessarily take multiple years to have material impact.

Environmental activism has been successful in closing coal-based electricity-generating plants, particularly older ones. This capacity cannot be replaced by renewable sources such as wind and solar in the short term because they are still generally too costly. I subscribe to the view that natural gas ought to be used as a bridging fuel until renewables are economically viable for base load. However, in chapter 18, "Will Cheap Natural Gas Hurt Renewables?" I discuss the belief that low gas prices will hurt the cause of renewables. This could change with price rises caused by increases in demand. The circumstances that could cause increases in demand are discussed in several chapters.

The book is laid out in sections with fairly self-contained short chapters on each topic. The intent is to allow the reader to skip to any chapter with a reasonable assurance of comprehension of the issues presented. But inevitably the understanding of some detail may well require that another chapter be read; in these cases references are made to the chapters in question.

While there is a solutions-oriented flavor to the environmental issues throughout, there is little doubt that this will be a tough row to hoe. It will need a combination of legislative oversight, technical innovation, and industry doing the right thing, with the last being nudged along by informed activism in the communities in which they (will) operate. The triple-bottom-line tenets of sustainable energy production must be put into real practice. In plain words: without profit there is no enterprise, but it must not be at the expense of either the obligation to environmental stewardship or the interests of the local community.

Shale Gas Basics

So, Where Did All This Gas Come From Suddenly?

"Do you believe in magic?"

—From "Do You Believe in Magic?" by Lovin' Spoonful (written by John Sebastian)

Few will dispute that shale gas has changed the very makeup of the petroleum industry. At every twist and turn new resource estimates appear, each vastly greater than the previous. The estimate in 2008 exceeded the one from 2006 by 38 percent. As with all resource estimates, be they for rare earth metals or gas, disputes abound. But through all the murk is the inescapable fact: there certainly is a lot of the stuff. How could this suddenly be so? The last such momentous fossil fuel find in North America was the discovery of Alaskan oil. But a discovery out in the far reaches of the US is understandable. In the case of shale gas all this is happening literally in our backyard.

To appreciate the excitement at the discovery of shale gas we first need to understand how oil and gas are formed and recovered. Millions of years ago, marine organisms perished in layers of sediment comprising largely silt and clay. Over time, additional layers were deposited and the organic matter comprising the animals and vegetation was subjected to heat and pressure. This converted the matter into immature oil, known as *kerogen*.

Further burial and temperature rise continued the transformation of kerogen to oil. The most thermally mature final form is methane, formed by the thermal cracking of oil. By and large the only real difference between oil and gas is the size of the molecule. Methane is the smallest, with just one carbon atom. One of the lightest oil components, gasoline, averages about eight carbon atoms. Diesel averages about 12. So, although we refer to them as oil and gas, chemically they are part of a continuum, so it is easy to understand that they come from a single source.

Oil Shale

Kerogen left in the original immature form is also found in certain shale deposits. These shale deposits are known as *oil shale*. The nomenclature can be confusing because a mere switching of the two words leads to a completely different product. (The oil found in shale is known as *shale oil* and is light oil in mature form ready to be distilled into useful products such as gasoline.)

Oil shale resources are estimated to be very large, greater than all conventional oil combined. The US has the largest deposits, primarily located on federal lands in Colorado, Utah, and Wyoming. However, the difficulty of extraction and processing has made this largely academic except in a few countries. Estonia is by far the most prominent user of oil shale, using it for electricity generation.

The immature kerogen requires processing to simulate what millions of years of heat and pressure accomplished in nature. Many of the deposits are very shallow and so can be mined. The mined material can be thermally treated to produce a hydrocarbon vapor, which when condensed yields a useful oil. By and large this is considered too expensive today.

Royal Dutch Shell conducted considerable research into heating the kerogen in place using a pattern of resistance heating elements. This process was piloted in Colorado but eventually mothballed. In principle, this in situ conversion is an elegant solution and likely the way forward in the future.

Petroleum

Petroleum principally comprises long-chain molecules with the formula C_nH_{2n+2}. In its simplest form $n = 1$, and this is methane, which is a gas at normal temperatures. When $n = 2$, this is ethane, also a gas, and one which features in later chapters. Propane has $n = 3$ and is a liquid at moderate pressure. Propane is most familiar as a cooking and heating fluid. Crude oil generally comprises long-chain molecules with n greater than about 20.

The greater the number n, the heavier the oil, because longer chain molecules are more dense. Refining is the process of breaking the molecules down (*cracking*) and often adding hydrogen as well, known as *hydrothermal cracking*. When all the oil produced was pretty light, with low n's, all one had to do was to heat it up. The different molecules would be separated based on their condensation temperature. This is known as *fractional distillation*.

The key word is *source*. The rock in which the oil and/or gas originally formed is known as source rock. In Figure 1, the source rock is gas-rich shale. The source rock is almost always shale, which is some mixture of silt and clay and sometimes some carbonates. Conventionally, the fluid in this rock migrates to a more porous body.

Figure 1. Schematic of the geology of natural gas resources

Source: US Energy Information Administration, January 27, 2012

The porous body to which the fluid migrates is depicted in Figure 1 as the sandstone, which is predominantly silica, an oxide of silicon. It may also be a carbonate, predominantly calcium carbonate. These two minerals are host to just about every conventional reservoir fluid in the world. The fluid (and by the way, gas is a fluid, although not a liquid) migrates "updip" (up the slope of the rock formation), as shown at the upper left and right. This is because the hydrocarbon is less dense than the water-saturated rock and essentially floats up, not unlike the oily sheen on your cup of coffee. This migration continues until it is stopped by a layer of rock through which fluid does not easily permeate. This is known as a *seal*, or more colloquially, a cap rock. Ironically this is most usually a shale, not unlike where the fluid (gas) originated. The trapped fluid is then tapped for production.

The trap is often in the shape of a dome, as shown in the upper left. It can also be a fault. This is when earth movements cause a portion of the formation to break away and either rise or fall relative to the mating part from which it just separated. In some instances a porous fluid-filled rock will butt up against an impermeable one, and a seal is formed laterally.

In Figure 2, the lightly shaded zone represents the sandstone, and the fluid of interest, shown in the triangular hashed area, abuts the cap rock, an impermeable zone shown in dark gray. This sort of faulting happened a long time ago, so the fault has healed. There is no fluid escape path along the fault.

Figure 2. Schematic of a fault with trapped fluid

In the early days of prospecting, explorationists looked for surface topography indicative of a dome-type trap below. Nowadays sound waves are sent down into the formation, using small explosions in most cases. The sound waves are reflected back, and an analysis of the various reflections produces excellent images of the subsurface. This is fairly similar to the ultrasound imaging of the internal organs of a human body for diagnostic purposes.

Unconventional Gas

I have described how conventional gas, and oil for that matter, are found and produced. The current flurry of activity in shale gas is concerned with going directly to the source. This was previously considered impractical, primarily because the rock has very poor permeability, which is the ease with which fluid will flow in the rock. The permeability of shale is about a million times worse than that of conventional gas reservoir rock. In fact, as I observed earlier, shale acts as a seal for conventional reservoirs. The breakthrough was the use of hydraulic fracturing. In hydraulic fracturing, water is pumped below ground at high pressures, causing a system of fractures in the rock into which the fluid is injected. These fractures are then propped open with some ceramic material, most commonly sand. Without this the sheer weight of the thousands of feet of rock above would close the cracks. The propped open fractures now constitute a network of artificially induced permeability, allowing the gas to flow out of the formation and up the borehole to be collected (in industry parlance, to be "produced"). This is akin to the use of pillars and beams in underground coal mines, in that case allowing passage of people and bins hauling coal.

The sheer ability to extract gas from source rock is now well understood as feasible. But some still doubt the magnitude of the estimated resource. Here is the explanation of why one would expect this resource to be plentiful. Consider that for a conventional reservoir to be formed, two events had to occur. First, there needed to be a proximal porous and permeable rock, and second, a trap mechanism had to exist. It would be easy to believe that more source rock did not have these conditions than did. In other words, the probability that source rock does not have a way to permeate to more porous rock surrounding it is greater than the probability that it does have such a release mechanism. This is why it is a reasonable conjecture that the total oil or gas trapped in source rock is greater than the amount of it that escaped into permeable trapped rock such as sandstone. Further adding to the potential is that this is fresh territory, relatively unexploited. Decades of exploitation have denuded conventional reserves, while the source rock further below remains relatively untapped.

A word on the nomenclature of resource estimation. A *resource* estimate indicates the quantity of estimated hydrocarbon accumulation, whether economically recoverable or not. A subset of that is a *reserves* estimate. Reserves are the portion of the resource that one could recover economically and bring to market. Typically in a new play (the term oil folks use for an active prospect) one would expect reserves to keep getting revised upward. This is because every new well put on production increases the certainty of the extent and quality of the reservoir, and the reserves can confidently be increased. In reading the popular literature it would be well to keep the distinctions in mind; they are often confused.

All of the press has been about North American activity. But once the basics are understood one can readily believe that source rock will be ubiquitous. The US Energy Information Administration (EIA) published a report (EIA, April 5, 2011) in mid-2011 providing its estimates of resources worldwide. China was seen with the most, followed by the US and then Argentina. More to the point, EIA estimates that source rock occurs worldwide, as I would have conjectured. The greatest impact will be on countries which are net importers, such as Poland, South Africa, Turkey, and the Ukraine. Poland is already aggressively pursuing development of the resource base. The have-nots of the hydrocarbon world can make some noise now.

The Oil Plateau and the Precipice Beyond

"Up, up, and away"
—From "Up, Up, and Away" by The Fifth Dimension (written by Jimmy Webb)

Christophe de Margerie, the CEO of Total, based in France, made a startling pronouncement in the fall of 2007. He said that oil production would soon plateau, at the level of 100 million barrels per day (MM bpd). Needless to say, it made quite a splash. Here was the CEO of one of the top five oil companies in the world saying there's a plateau coming. At that time the world was producing about 85 MM bpd and consuming the same amount. In fact, that has been the pattern: consumption equals production, and price modulates demand.

After that I personally, publicly asked a CEO of a major oil company to comment on de Margerie's prediction. He acknowledged the plateau as being real. He said, "I'm not sure I'm going to subscribe to the 100 number, but there's a plateau coming." Shortly before that I spoke to the head of the French Petroleum Institute (IFP), who confirmed that their modeling showed the same thing. They pegged it at a somewhat lower number.

So here we have substantial people saying there's a plateau coming and yet nobody acknowledges it publicly. Nobody wants to discuss it. Nobody really wants to act on it.

The significance is that if this is supportable, and a case can be made for recovery-driven economic growth, a supply/demand imbalance will cause havoc. The solution in part is to find substitutes for oil. Cheap natural gas can be one avenue. Hence the inclusion of the oil plateau discussion in this volume. But first the thesis.

Causes

Now you'll ask the reasons for the plateau. First of all there is a technical model that predicts a plateau, courtesy PFC Energy in Washington, DC, which I discuss to some degree below (PFC Energy, 2009). But some qualitative arguments can be made in support of a plateau hypothesis. For example, national oil companies have realized they have a resource they need to husband. International oil companies used to move into resource rich countries and extract oil via production sharing contracts. Such contracts had built in the incentive to get the most oil out as quickly as possible in part because the contracts had limited duration.

There's a truism in oil and gas production: if you draw it out quickly, then the net recovery—that is, the fraction of fluid in the reservoir that is ever recovered—reduces. The industry currently leaves behind about two-thirds of the oil in place. When the international oil companies went into these resource-rich nations, they drew oil out as quickly as they could because that maximized the value of the contracts. That was not in the best interest of the national resource.

Increasingly the nations with oil resources have figured that out. Now they are forcing the issue, telling the international oil companies, "We'll do it ourselves. We don't need you." The key point is they want to bleed the oil out in more measured fashion. Guess what that does to production rates?

Most of the major oil companies are therefore forced to seek unconventional sources of oil—for example, Canada's Tar Sands—which are largely heavy oil. Additionally, Alberta now has a small carbon tax on oil from the Tar Sands.

The late Matt Simmons, a highly respected figure in oil and gas investment circles, wrote that Saudi Arabia would not be able to open the spigots—that they didn't have the oil (Simmons, 2005).

Predictably, the Saudis have been mum on the point. The fact of the matter probably is that the Saudis have the oil, but they've got a different view of it now and how to release it. The national oil company Aramco has been a leader in the application of technologies to maximize recoveries and so can be expected to add to the nation's productive capability. But they're not going to get bullied into releasing it faster just because the world wants a lower price of oil. People thought of Saudi Arabia as the buffer, that they'd just open the dams, but it just doesn't seem like they will. Matt Simmons took the position that they could not. It's irrelevant: they won't. Whether they can't or won't, the result is the same: they will no longer make up shortfalls elsewhere unless driven by political exigency.

Consumption versus Production

The estimated production plateau of 95 MM bpd—I think PFC at this point is talking about 90–92 MM bpd—comes dangerously close to the 87 MM bpd we were consuming prior to the recession. Today that consumption (demand) is still down around 85 MM bpd.

Figure 3 is derived from data from PFC Energy. Their demand curve crosses over the supply around 2020. This is a scant eight years away. PFC makes the point, as do I, that the only real solution is finding alternatives to oil utilization. Note that they show the plateau for only few years, and then a decline. All of this is subject to speculation. But the key takeaways are as follows: more oil has to be found and produced, and alternatives to oil urgently need to be developed. The urgency draws from the fact that capacity ramp-up of biofuels and the like takes billions of dollars and multiple years to take hold.

Figure 3. World oil supply versus demand projection

Source: Adapted from PFC Energy data (PFC Energy, 2009)

The key factor is the speed of the recovery with respect to automotive use. In the United States at least, oil is about transportation. Gas is about power and petrochemicals. The plateau is real and the recovery is real. It's very real in China and India, which never really saw much of a recession. They each have close to double digit growth in GDP for the last several years and continue apace. In China and India, what do you think newly prosperous people do? They buy a vehicle. They go from a bicycle to a motorcycle or scooter to a car.

Shown in Figure 4 is per capita car ownership on the vertical axis as a function of per capita GDP. China is near the bottom, and India is even further below, almost off the graph. These are countries with the fastest growth. The US is on the upper right, the direction in which these others are headed.

Figure 4. Per capita car ownership as function of per capita gross domestic product (GDP)

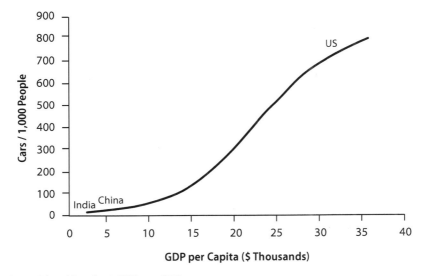

Source: Adapted from data in PFC Energy, 2009

All of this indicates that transport fuel usage is likely to keep increasing, and that if it does, the crossover point between consumption and production is probably sooner than later (I'm not talking electricity—that's a completely different argument).

Any target for reduction in imported oil requires displacement of oil-based transport fuels. The rise of wind- and solar-based electricity cannot be juxtaposed with oil usage reduction. This is because very little transport runs on electricity. Years from now, when electric vehicles are a significant fraction of active automobiles, that could change.

The plateau is coming, and if consumption continues at the current rate, a crossover is coming, too. And at the point of the crossover, we're not talking a spike in prices. We're talking a sustained price increase. A spike is driven by a shortage at some point. This is not a shortage at some point. This is a plateau. Clearly demand destruction will occur in response to this. But the initial surge in price will be damaging.

Here is the crux, though: Do you really want to test the plateau theory? The alternative to testing it is doing something smart, like replacing oil with something that is more environmentally responsible.

The Role of Shale Gas

Prior to the intruding reality of massive amounts of cheap natural gas, the options for oil substitution were limited. Electrification of transportation was, and remains, the most important avenue. However, by its very nature, the carbon footprint burden is merely shifted elsewhere. The electricity generation process generates carbon dioxide, even though the vehicle emits none. But certainly oil would be saved. The Fukushima Daiichi nuclear disaster placed a damper on nuclear energy as a source of clean electricity. Germany placed a moratorium on new plants, as did Switzerland, and the raising of the risk bar has caused funding to dry up in other countries. That pretty much left coal, natural gas, and renewables.

Clean coal is simply not going to have legs until we have a price on carbon, be it a tax or some other measurable way to pay for cleaning it up. Is it technically feasible? Yes. But it will add in the vicinity of 3–5 cents per kilowatt-hour to clean it to natural gas levels. Aging plants will be loath to add new equipment no matter what the incentive.

Wind and solar will continue to get increasingly viable. However, they simply will not make a major dent for decades. Also, to be viable alternatives, they need a price on carbon (in the form of a tax or penalty associated with carbon emissions) or policy boosts. The solar subsidy in Germany appeared to work but the post-subsidy world will be rough.

That leaves natural gas. Shale gas is causing the price to remain low in North America. There is reason to believe that similar effects will be felt elsewhere. A dramatic shift from coal to natural gas for electricity is likely and feasible.

Shale Oil and Natural Gas Liquids

This is a huge wild card in the oil game. First some definitions are in order. Shale oil is oil found in shale, much as the gas is. This is distinct from oil shale, which is sedimentary rock containing an immature form of oil that needs significant processing. It is generally believed that the oil shale resource, also vast, is not feasible for exploitation in the foreseeable future. North America is especially blessed in this regard, but technical breakthroughs are needed to economically "cook" it in place to make its use viable.

Shale oil is found in two settings. One is in the oil portions of shale gas reservoirs, and the other is in completely different areas with no appreciable associated gas. The outstanding example of that is the Bakken formation in North Dakota, Wyoming, and Montana and extending into Canada. As in the case of shale gas, fracturing is necessary to release the liquid. The production rates are modest per well. But the production is economical and the resource estimates are huge. Unlike that other unconventional oil, heavy oil, shale oil is light and sweet, meaning it is low in sulfur, and the size of the molecules is on average small. These characteristics by and large make it a refinery darling because it is less costly to convert to gasoline.

Assuming production from shale oil is substantial, how does one rationalize the existence of an important resource against the plateau effect? For one thing, the plateau is likely driven by lack of access. The US, as a net importing nation and one not limiting access to at least this type of resource, is positioned to become an important new producer. As a consumer of about a fourth of all oil produced, significant domestic oil production will eventually affect world oil prices. But that is not on the medium-term horizon. The very nature of the oil shale resource limits the growth ramp. But it could materially affect the supply/demand equation.

Natural gas liquids (NGLs) are liquids found in association with shale gas. Because the price of gas is low, and will likely remain low for years, most of the activity is likely to be in the portions containing NGLs, known as *wet gas*. These are portions of the reservoir with a higher proportion of the liquids. The liquid price is pegged to oil prices, not gas. If oil remains tight in supply, as hypothesized above, the price can be expected to remain high. Today oil is roughly four times the price of gas on the basis of energy content. One can expect that ratio not to get better and possibly to get much worse in the coming years. So wet gas is where the action is and will stay until gas prices rise. That could happen with demand improvement. But wet gas will always be more profitable.

Wet gas will have liquids associated with natural gas in quantities ranging from 4 to 12 gallons per thousand cubic feet (mcf). In the Marcellus Shale, it averages 7 gallons per mcf. That is 0.17 barrels, since each barrel is defined as

42 US gallons. Take an oil price of $100 per barrel, essentially what it is today. For natural gas we will assume a price of $4 per mcf, even though at the time of this writing it is much lower due to a warm US winter. The liquid component is worth $0.17 \times \$100 \times$ a discount factor added for conservativism. Even taking the discount parameter to be 0.3, the liquids are worth $5.10, while the associated natural gas is worth $4. No matter what reasonable discount you apply, the liquids materially add to the profitability of the gas. Small wonder that wet gas is the play.

In chapter 12 I discuss the ethane dilemma. One reason the discount factor is low is that ethane, which can constitute over half the liquid (actually ethane is a gas in that state, but it is lumped in with the NGL by the industry), is priced half that of the bigger molecules such as propane and butane at this time. Ways to monetize this are discussed in that chapter.

Gas Will Remain Cheap and Displace Coal

"She got the gold mine, I got the shaft"
—From "She Got the Gold Mine" by Jerry Reed (written by Tim DuBois)

Natural gas will rapidly displace coal for power even without the benefit of a price on carbon.

When natural gas is combusted for power production, the carbon dioxide produced is about half that produced from coal. Beyond this, the externalities associated are small when compared to those attributed to coal consumption, notably the possibility of emissions of mercury, sulfur, and NOx (with important environmental and health impacts) and less noticeably fly ash disposal problems. A decided shift toward natural gas in the early years of this century stalled when gas prices became unpredictable. The shift started when gas was at about $2 per million BTU (MMBTU) and the all-in cost of electricity production was less than with coal. A Bernstein report (Wynne, Broquin, & Singh, 2010) details the history of this, including the quandary reached in 2007 when gas prices were going up but so were construction costs for coal plants. This caused a large number of planned coal plants to be postponed.

Today Congress has reached no consensus on legislation to mitigate carbon emissions, and the political makeup of the current Congress makes that prospect less likely. So, a price on carbon is unlikely in the near future. Mind you, the European version of cap and trade has simply not worked. For years the price has fluctuated from about €15 ($19.50) per metric ton (tonne) to about €25 ($32.50). In late December 2011 it actually crashed to €6.50. Uncertainty in price dampens investment because there is no choice but to make the discount rates higher. That raises the overall cost of investment. A price on carbon much less than about $40 per tonne will have little effect because of the cost of sequestration. In Europe one reason for the low prices has been the very high allowances given to the major emitters. Even if a cap and trade scheme were to be devised in the US, there is every reason to believe that a deeply divided Congress would offer similar generous allowances.

However, the US Environmental Protection Agency (EPA) is expected to levy strong restrictions on mercury, sulfur, and NOx. One report (Wynne et al., 2010) states that more than 40 percent of coal generation plants that do not meet these standards are over 50 years old. One can reasonably expect many of these to be mothballed, leaving room for new coal plants, gas-fired plants, or alternative methods such as nuclear, with the problem exacerbated as the global economy continues to expand. With gas plants costing less to build, being cheaper to operate (at forecast gas prices), and easier to obtain regulatory approvals for new construction, the bias toward gas power generation will be significant. The Fukushima Daiichi nuclear plant disaster in Japan is having a chilling effect on the nuclear option. Switzerland banned the building of new nuclear plant reactors, and Germany went back on its agreement to extend the life of existing plants. Some of this action occurred within days of the tsunami and subsequent reactor meltdowns and no firm finding regarding the true impact. Nobody is seriously considering wind or solar for base load any time soon.

The lead time to first electricity production is low for gas, as compared with coal or nuclear. Coal and gas are polar opposites on components of cost. For coal the capital cost is a high proportion of the all-in cost, around 60 percent, whereas the commodity is relatively stable. In the year 2010, coal varied little from a cost of about \$2.25 per MMBTU. This is the best way to express the cost of coal because the BTU content is highly variable; from brown coal to anthracite, the BTU content differs by more than a factor of 2. So in comparing coal with other fuels, the price per ton is not a meaningful figure, even though oft quoted in reports. BTU content is a better measure of comparison.

Gas-based power, on the other hand, has a relatively low capital cost, around 15 percent of the all-in cost, and in the past a highly variable price, at least over the lifetime of the amortization of equipment. Consequently, while coal power is highly susceptible to construction costs and inflation in general, gas economics depend primarily upon the price of the commodity.

Figure 5 plots the all-in cost for electricity produced from natural gas as a function of gas price. The plot uses data from the Bernstein research report cited above (Wynne et al., 2007) and reflects 2006 economics. The price in October 2010 is the most recently available monthly average and is shown on the figure. It was episodically over $5 per MMBTU during the year and also below $4. The generally accepted cost of production from new coal-based construction meeting emission standards is drawn as a horizontal band between 6 and 6.5 cents.

The impact of a price on carbon can be treated in one of two ways. Post-combustion carbon capture using technology known today, but not yet perfected, can be expected to cost about 3 to 5 cents per kWh to bring the carbon dioxide emissions down to natural gas carbon emission levels. This is plotted as the second horizontal band above the first, plotted using the more conservative figure of 3 cents. The second way to treat the impact of a price on carbon is to simply pay the carbon penalty. At a price of $25 per tonne this will add around 1 cent. This accentuates the point that any price less than about $40 per tonne is simply not reflective of the cost of taking the carbon out.

Figure 5. Comparison of electricity cost: coal versus natural gas, as function of gas price

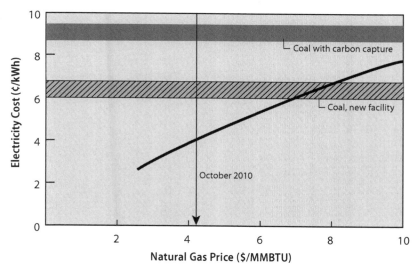

Source: Wynne, Broquin, & Singh, 2010

Of immediate note is the fact that in late 2010 the gas-based cost was well below that from a new coal plant. Of particular note is the observation that the breakeven with coal kicks in at a natural gas price between $7 and $8 per MMBTU. *This is without consideration for any carbon penalty.* If that were to be included, the breakeven moves to gas prices between about $9.50 and $12 per

MMBTU, depending on which model of carbon penalty one uses. The gas price spikes of several years ago still endure in memory. The fact is that prices were over $12 for only about four noncontiguous months. Predictability of price and assurance of supply are needed for a major shift from coal to gas.

Assurance of Supply

A scant three years ago this issue would have had a different answer. We would have been discussing liquefied natural gas (LNG) imports and environmental risk posed by that. Shale gas has changed all of that. The US can expect to be self-sufficient for a hundred years. In fact, it may well become a net exporter of gas or gas derivatives.

This could also change the entire debate around natural gas in Prudhoe Bay, Alaska. Natural gas in copious quantity has been reinjected because of the high cost of a pipeline to the Lower 48. Under the circumstances a substantial bullet got dodged by the delay in that decision. Had a pipeline been built, the fully loaded cost of the gas would have been challenged by that of shale gas production and the utilization could have plummeted. As an alternative, the nation should reopen the possibility of exporting Alaskan North Slope LNG. Due to the net shortage in the US this had been politically impossible except for a single permit for LNG export of gas from Cook Inlet. That should no longer be an issue, and in fact ConocoPhillips has sought an extension to the Cook Inlet permit. Shale gas from the Horn River Basin in British Columbia is already slated for LNG export by Apache. Gas and gas derivative export from North America could become a trading force.

Gas Prices in the Future

The floor will be determined by demand. In 2010 it hovered around $4 per MMBTU even without demand creation. A shift from coal to gas will drive demand. So the concern is likely more at the upper end. As I have discussed elsewhere, assurance of a moderate price is a huge driver in the chemical industry. Is there a mechanism for a ceiling? The answer is affirmative and results uniquely from the setting in which shale gas is found.

Most shale gas is either proximal to the intended market, as in the case of the Marcellus, or close to major pipelines, as in the case of the Barnett (Texas), Haynesville (Louisiana), and Woodford (Oklahoma) shales, to name just three big ones.

Compared to conventional gas, these wells are relatively shallow and on land. In a study conducted about a year ago (Nikhanj & Jamal, 2010), examination of 100,000 wells showed that the average time for the construction of a horizontal producing well from the beginning of a well to the start of the next one was 27 days. With the steep learning curve to which this industry is accustomed, these numbers have doubtless improved even in one year. In fact, shale gas is particularly advantaged in this respect because the sheer number of wells allows for innovation to be perfected in what amounts to a factory situation. However, the time from the start of drilling to the point of delivery to sales has been much greater, averaging closer to 150 days. This does vary by area. The Haynesville Shale, with the longest drilling times due to the depth and complexity, curiously has the shortest time to market, about 90 days.

Pad Drilling

The concept of drilling up to 15 or more wells from a single location (pad) was a technique pioneered in Colorado as an environmentally friendly technique to minimize roads, among other features. However, this may be the culprit for the long time to market, because it allows wells to be drilled and completed in batches. Batch drilling is a technique wherein the early portions of a number of wells are completed at the same time, followed by the later portion of the wells all together. This reduces overall cost in many instances. But it necessarily greatly increases the time from first well start to first fluid delivery.

That said, the pad technique is likely here to stay, particularly in the heavily populated areas of the Marcellus. It dramatically reduces the traffic associated with pump trucks, proppant delivery, water disposal, and the like. The aggregation also permits higher levels of sophistication in areas such as water handling, remote decision making, and quality control. Ultimately operators will balance the economics of batch operations against the opportunity cost of later sales.

So, depending upon the area and the business drivers, new production can be brought to bear in as little as 90 days and certainly under 180 days. This is compared to over four years for a conventional offshore gas field. This short time span will basically keep a lid on the high end. Speculators will be aware of the quick response ability. This ability to service demand by bringing on capacity at short notice will likely keep prices under $8 per MMBTU, possibly closer to $6 per MMBTU. This is the crux of the thesis that shale gas will enable natural gas to be in a tight band between $4 and $6.50 per MMBTU for a long time.

At numbers north of $4 per MMBTU most operators will make a very good profit, especially if a portion of the gas is laden with natural gas liquids.

Newer technologies will continue to drive down cost, as has been the case in every new resource play since Colonel Edwin L. Drake's historic oil find in Pennsylvania in 1859. So the businesses will be sustainable and continuous supply assured. Note: LNG has a built-in cost of nearly $3–4 per MMBTU just to liquefy, transport, and re-gas, over and above the gas production cost.

In summary, shale gas has changed the game for all industry and power in particular. For the first time in memory, a major fuel can be predicted to be priced in a tight band at low to moderate levels for many years. This is the type of certainty that drives investment. The timing of the realization is propitious in dealing with upcoming new EPA regulations on coal emissions and possible regulations on CO_2. No new coal generation plants are economically justifiable provided our predictions on natural gas pricing hold up.

What a Difference a Hundred Million Years Makes

"Time is on my side"

—From "Time Is on My Side" by The Rolling Stones (written by Jerry Ragovoy)

Utica, New York, is an unpretentious city of 60,000 people that has an interesting geological feature. Here exists a rock outcrop which is named after the city: the Utica Shale. This shale body is just beginning to make waves in the natural gas industry. This rock is old—100 million years older than the Marcellus Shale, the current darling of the shale gas industry. The operative word is *current*. The Utica appears set to upstage the Marcellus.

The Utica Shale is directly below and greater in lateral extent than the Marcellus. If all of it proves productive, it will likely be the largest gas field in the world. If the initial production numbers prove typical, the Utica may be one of largest in reserves as well. With the Marcellus itself no slouch, the two together represent an immense concentration of hydrocarbons. The Utica Shale extends into the Great Lakes. This offers the possibility of offshore shale gas operations, which would be a first.

The Utica is generally thicker than the Marcellus. In the primary regions of interest it appears to vary from 150 to 500 feet. Somewhat complicating the issue is that in some areas the fluid had migrated to sandstones and carbonates immediately above and adjacent to the shale, so the source rock may not be the only producer, unlike the Marcellus. The percent total organic content (TOC) is the amount of organic matter present and is a measure of the economic potential of the resource. This is somewhat lower in the Utica than in the Marcellus.

This difference in TOC is one of the reasons for caution regarding *resource* estimates as compared to *reserve* numbers. The latter value is always smaller for any prospect because to be qualified for a reserve designation, the fluid has to be economically recoverable and deliverable. In general, a TOC less than 2 percent is probably below the threshold of interest.

Productivity Considerations

Once the TOC threshold is crossed, the next consideration is the estimate of the gas in place. In conventional reservoirs, the gas is all free gas in the pore spaces. It is then driven out by a variety of drive mechanisms. Shale layers have two types of gas. The first is free gas in the pores. The other is gas adsorbed onto the organic matter. *Adsorption* is a surface phenomenon, as opposed to *absorption*, which is a bulk effect. An everyday example of adsorption is the activated carbon filter used in water filter systems to remove select harmful species such as bacteria. Activation is a process that dramatically increases the surface area of the carbon and increases its adsorptive ability. In the case of a water filter, the harmful species stay adsorbed on the carbon and the filter element is replaced from time to time.

In the case of shale gas, the adsorbed gas is released simply by reducing the pressure by opening up the well to the surface, and further reducing it through removal of the free gas. In principle one could improve desorption (the release of adsorbed gas) through injection of carbon dioxide. This molecule preferentially adsorbs on shale organic matter by about a factor of 5. Thus all the adsorbed methane would get released; this would be a CO_2 sequestration mechanism. The practicality of conducting this operation needs to be researched.

Shale usually comprises a combination of silica and clay, often with another mineral such as calcite or calcium carbonate. The relative quantities determine the ability to produce and propagate fractures. The best rock for successful fracturing is slightly brittle. Too much clay makes it too ductile. Think peanut butter as opposed to peanut brittle.

What Makes Utica Different

The source rock belongs to the Ordovician Period, which extends to 500 million years ago and which is about as old as rocks derived from sediments can be. As a frame of reference, dinosaurs started 250 million years ago and got erased 65 million years ago. The black shale of the Utica is about 460 million years old, as compared to the Marcellus, which is around 370 million years old. This extra age manifests in two ways. One is that it is deeper wherever they overlay each other. This difference is up to 7,000 feet. The extra depth brings with it higher natural pressures for fluid production, which would be more similar to the prolific Haynesville Shale in western Louisiana and eastern Texas. Of course the deeper wells will be more expensive.

The extra age also brings with it a greater thermal maturity. This means a greater conversion of kerogen to oil and gas. All of this is good for productivity. In addition, the close proximity to sandstone and carbonate layers causes some carbonate content in the shale, making it more porous. This offsets the lesser TOCs compared to the Marcellus. But the carbonate content makes the formation more like the Eagle Ford Shale in Texas than the Marcellus in its fracturing character. So, despite the fact that Marcellus property leasers have the Utica below them, the techniques used would need Eagle Ford–type experience.

Producing the Utica

A point being debated is whether an operator could exploit both reservoirs at the same time. On the plus side, the drilling and fluid export infrastructure certainly could be used for both. On the other hand, with the Utica being at much higher pressures, it would be impractical to mingle Marcellus and Utica fluid. Expensive techniques known as smart wells could be employed, but the more likely option would be to produce them sequentially. Utica would go first and later the vertical portion of the same well bore could possibly be used for the Marcellus.

Another distinguishing feature of the Utica is the presence of a significant oil portion on the western side. Much of this will be in Ohio, a relative newcomer to the shale bonanza and one in dire need of job creation. In fact the field it is most similar to is the Eagle Ford Shale. It has the same pattern of hydrocarbon distribution: oil, wet gas and dry gas. The Eagle Ford has been the fastest growing shale area by far as measured by the rate of increase of production; one could likely expect the same with the Utica. Chesapeake Energy has recently reported results from wells in Ohio and they are huge. Gas production was stated to be between 3 and 9 million cubic feet per day and with associated natural gas liquids (NGL) of 800 to 1,500 barrels per day. The ratio of liquids to gas is similar to the numbers in the Marcellus, but the raw numbers are much greater, as would be expected from the greater depths.

The Oil/Gas Price Spread and the Effect

A scant half decade ago, oil and gas prices were in substantial parity. A barrel of oil has about six times the energy content as a thousand cubic feet of natural gas. So, if the prices are in parity, the ratio of price of oil to natural gas should be around 6 to 1. The last sustained period for that price ratio was back before 2007. The trend since then has been for the ratio to be in the mid-20s to 1.

Beyond Gasland

"We can work it out"
—From "We Can Work It Out" by The Beatles (written by John Lennon and Paul McCartney)

No shale gas production issue may be more fraught with partisan rhetoric than the possibility of methane in drinking water. Flaming faucets make for great imagery no matter the true frequency of occurrence.

Well water contamination is very personal and frightening. Think Erin Brockovich. Airborne species don't appear to get the same reaction. Certainly, carbon dioxide in the air barely registers on the average personal anxiety scale. Consequently, assaults on the quality of well water make for avid reading and activism. In the case of shale gas, industry response to well contamination has been sweeping in denial. Both sides play fast and loose with the English language, as will be shown.

There are two potential ways in which shale gas operations could contaminate aquifers. One is through leakage of the chemicals used in fracturing. These would then be liquid contaminants. The second is the infiltration of aquifers by produced methane. This is a gaseous contaminant, albeit in the main dissolved in the water. If present, a portion may be released as a gas, as spectacularly depicted in the documentary *Gasland*. Natural occurrences such as the Eternal Flame Waterfall in the Shale Creek Preserve in New York demonstrate methane intrusion into a fresh water source. The name follows from the fact that the gas remains lit with a visible flame under the rock overhang of the waterfall.

Incidentally, methane is odorless and colorless, so it is hard to identify. It leaves no taste in the water, but can be hazardous if it collects in an enclosed space. When used in commerce, industry deliberately adds mercaptans, a smelly substance (travelers on the New Jersey Turnpike proximal to refineries in years past know the smell well) for added safety. Gas leaks in a commercially supplied setting such as a kitchen or furnace are therefore detectable simply by smell. Leaks from a well would not be detectable by smell.

Natural contamination is either from relatively shallow biogenic methane or from thermogenic gas from deep deposits escaping up along faults and fissures. These last are generally due to tectonic activity at some time. The two types of methane gas have fairly different fingerprints and can often be distinguished on that basis. Good oil and gas exploitation practitioners will avoid producing in areas with significant vertical leak paths because they vitiate normal sealing mechanisms.

Distinguishing Between Methane Sources

Thermogenic methane is formed when heat and pressure acts upon organic matter present in the sediment. The stress of this "thermal maturation" causes the longer molecules to be chopped up to form smaller molecules. The smallest is methane and usually dominates the mix. Also present will be the somewhat larger molecules such as ethane and propane, collectively known as natural gas liquids (NGLs). Thermogenic methane will almost always have some of the larger species present.

By contrast, biogenic methane will almost never have these other gases. This is because the method of formation is bacterial action upon carbon-bearing molecules such as carbon dioxide and hydrogen, yielding methane molecules, water, and energy. This mechanism is unlikely to form bigger molecules.

The absence of larger molecules is generally a clear indicator of biogenic origin. The reverse is not necessarily as clean an argument in part because mixtures of gases from different origins are possible.

Another distinguishing feature is the isotope signature. The scientifically faint of heart can skip this paragraph without loss of the thread. Carbon has a heavy isotope ^{13}C, with an extra neutron compared to the normal species. When methane is formed by either mechanism above, the carbon atoms in the methane preferentially have a higher proportion of light atoms compared to the source material. This is known as fractionation in that the heavy atoms are preferentially left behind. But bacteria tend to want to feed on lighter carbon and so the fractionation is more pronounced than it is for thermogenic reactions. Also, the stuff they feed on is often of bacterial origin so is already concentrated with the light version. So, the fractionation is an indication of origin.

The isotope signature, taken together with the level of presence of larger molecules, is the generally accepted means for distinguishing between gas origins.

Despite the seemingly sound scientific fingerprinting techniques available, identifying the source of gas is subject to interpretation. This is particularly the case when, within the class of thermogenic origin, efforts are made to identify the age of the rock from which the leakage took place. In principle, both the isotope signature and the NGL content have relationships with rock age. These relationships are not always well behaved enough to be unequivocal.

Water Well Contamination Studies

The first comprehensive study (Osborne, Vengosh, Warner, & Jackson, 2011) of possible contamination of water wells from shale gas production activity was by Osborne and colleagues from Duke University. They studied 60 wells in Pennsylvania and found no fracturing chemicals in the water wells. They did report finding a strong correlation between proximity of shale gas wells and methane in the well water. The data have been interpreted by the authors as proving the origin of the methane to be from the Marcellus formation from which the gas was being produced. The method used was as described in the box at left. Since no baseline testing of the water wells was done, the entire conclusion rests upon the method of analysis. Any retrospective study faces this shortcoming.

More recently, industry proponents took the same data and published an analysis to the effect that "the isotopic signatures of the Duke study's thermogenic methane samples were more consistent with those of shallower Upper and Middle Devonian deposits overlying the Marcellus Shale. This finding indicates that the methane samples analyzed in the Duke study could have originated entirely from shallower sources above the Marcellus that are not related to hydraulic fracturing activities"(Molofsky, Connor, Farhat, Wylie, & Wagner, 2011). They based this in part upon other studies they conducted in which they tested 1,700 wells. As a clarification of terminology, the geologic periods named above span about 30 million years—not huge in geologic age terms. That is why they were careful to use the conservative phrasing "could have" in relation to their assertion. They, too, did no baseline testing and note that the area is well known for methane intrusions in freshwater aquifers. They provide a detailed description of the geology of the area and identify many small gas bearing formations. In my recommendations below I suggest regulations requiring the identification and sealing of these gas-prone intervals in the shale gas wells.

Such dueling interpretations are inevitable when dealing with data with no baseline studies. The only rational way to determine the propensity for methane contamination attributable to gas production is to conduct baseline analyses of water wells prior to any activity. Sound inquiry in any field of science has always had this feature. A paper on biology experiments would be rejected in a peer-reviewed journal if these "controls" were not present. Both sides agree with this, and in fact the EPA is conducting such studies as part of the congressionally mandated investigation of fracturing operations. The most thorough type of study would be one that did all of the above and also considered the geology and the integrity of gas well execution.

The distinction between potential liquid and gaseous contamination is important because the hazards are different, as are the remedies and safeguards. Also, because well water could not naturally have the liquid contaminants from fracturing fluid, any presence at all is evidence of a manmade source. Therefore, simple testing of wells proximal to drilling operations is sufficient, with the only possible complication being some source other than drilling, such as agricultural runoff. This is easily resolved because of the specificity in the chemicals used for fracturing. The fact that Osborne et al. (2011) found no such chemicals in any wells is hardly ever reported in the press, indicative again of the polarization caused by the issues.

Unfortunately, liquid and gaseous contaminations get lumped together in the statements by shale gas opponents and also the genuinely concerned public. Some see methane intrusion as proof of well leakage as a whole and therefore equate it to chemical contamination as well. *Gasland* reports "thousands of toxic chemicals" as the hazard, which is hyperbole. In actuality, the mechanisms for possible leakage of gas and liquid are quite different. Methane as gas is much more likely to leak out of a badly constructed well than is a liquid.

So, do producing gas wells sometimes leak into freshwater aquifers? The answer is yes. In all cases this is because of some combination of not locating cement in the right places and of a poor cement job. Many wells will have intervals above the producing zone that are charged with gas, such as the formations identified in the *Oil and Gas Journal* paper cited above (Molofsky et al., 2011). If these are not sealed off with cement (many wells are not cemented top to bottom by design) some gas will intrude into the well bore. This will still be contained unless the cement up near the freshwater aquifers has poor integrity. In that case the gas will leak. You will notice nothing in the prior discussion says anything about fracturing. In other words, a badly constructed well is just that, no matter how the gas was released from the formation.

This distinction is lost on many. The paper by Osborne et al. (2011) unequivocally shows no fracture chemical intrusion into water wells. It also shows gas intrusion in disturbingly many cases, although no baseline measurements were made to normalize for possible natural seeps and prior drilling activity. Yet the title of the paper is "Methane Contamination of Drinking Water Accompanying Gas-Well Drilling **and Hydraulic Fracturing**" (emphasis added). The last three words infer a causality that is not proven and in fact is contraindicated by the absence of fracturing chemicals in the water wells.

Industry proponents on the other hand make statements such as "hydraulic fracturing has never contaminated groundwater." In precise terms this may be right in that fractures have not propagated into groundwater. Take the hypothetical case of a well associated with fracturing operations that leaks gas but not liquid. One could argue that the poor construction would simply not have occurred but for the desire to fracture the shale reservoir to produce the desired fluid. So an opponent would take those very data and say "hydraulically fractured wells contaminate groundwater," while the proponent could say "hydraulic fracturing did not contaminate groundwater." Neither would be wrong. It is the public that will be confused with this license taken with the language.

Suggested Remedies

Rhetoric aside, proper stewardship of our resources and the environment is possible. Some possible measures are listed here and policy suggestions are made in chapter 24, "Policy Directions." Permits must be given only to oil companies with good track records thus maximizing the chances of diligence in well construction. Water wells proximal to intended operations (the Pennsylvania governor's Marcellus Shale Gas Commission recommends 2,500 feet) should be tested prior to drilling, at the cost of the operator. The state Department of Environmental Protection should maintain a list of certified testing laboratories and only these ought to be used.

Logging ought to be required to identify zones of possible minor gas production. Logging is a routine technique for specialized sensors to be lowered into the well to identify formations of interest. The state department of environmental protection must require that these zones of gas be sealed with cement. Consideration may be given to requiring cementing top to bottom in areas whose geology dictates that. At a minimum, the cement bond log ought to be required. This examines the integrity of the cement job and famously was not run on the *Deepwater Horizon* well that blew up in the Gulf of Mexico.

Routine testing of the water wells ought to be de rigueur, with a prompt attempt to seal the well if leaking. This occurrence should also result in a severe penalty. Regulations requiring all of the above will not be considered punitive by the industry. This is because normal sound practice includes all of these, with the exception of the stipulations with regard to water well testing. That is not done routinely, although there are several instances of this happening voluntarily. All of this and adherence to sound drilling and completion practices are necessary to ensure the sustainable production of a valuable resource.

Preventing Contamination of Surface and Groundwater

"A bridge over troubled water"

—From "Bridge Over Troubled Water" by Simon and Garfunkel (written by Paul Simon)

The safe handling of water associated with shale gas operations is the most significant environmental hurdle faced by the industry. This can and must be surmounted. This is not to downplay fugitive methane emissions or any other concern, but this one is up close and personal to the community surrounding gas production. The global warming propensity of methane, and arguments regarding the different models to assess the impact—these simply do not have the resonance of potential short- and medium-term health effects of polluting surface or groundwater. These last are what one could call front-of-the-box items.

The potential for pollution of surface and groundwater comes from two sources. One is the improper handling of the return water from fracturing operations, known as flowback water. The other is from accidental spills of all sorts, including fracturing chemicals and water laden with these chemicals. The very nature of shale gas formations is such that even if fresh water is injected as the fracturing fluid, the returning water is much more salty. The salt content could be as much as 350,000 parts per million (ppm). Sea water is about a tenth as salty, and fresh water is defined as less than 500 ppm. The saltiness is due to the fact that the returning water includes water resident in the formation, which happens to be heavy brine.

The Front of the Box

A recent *New York Times* story (Kaufman, 2011) has a very interesting take on the environmental movement and changes therein. These organizations in the past have taken national or even global approaches to the issues. The rise of global ambient temperatures caused by greenhouse gases is a case in point.

The general public can be left cold at two levels. One is that global issues do not resonate with a lot of folks, while local ones do. The other is the discounting of future privation. This is not unlike discounting future earnings in finance; a discount rate is applied which gives a lower *present value*. Similarly, future suffering is discounted, especially when it is 40 years out, as are most global warming warnings. Rising water levels on a Florida beach 40 years hence (and only a maybe at that) have little resonance with the public in Wyoming. One could call it two degrees of separation.

The *Times* story draws a clever analogy. If a consumer is walking down a grocery store aisle and sees a box with a delectable brownie on the face, she may be attracted to it. Some might look at the back of the box, which details the information indicative of a potentially obese future for the consumer of the goods. Even though the future in this case is more in the short term than the aforementioned global warming one, the choice of looking at the back of the box—at the potential hazards—is personal and will not happen all the time.

The *Times* story concludes that environmental activism is best served by presenting front-of-the-box issues at a local level, and leave the back-of-the-box issues to the national or global level.

Radioactive Elements

Naturally occurring radioactive elements can be found in the flowback water. These can be isotopes of uranium, thorium, potassium, or radium. *All* shale strata have some amounts of radioactive species naturally. In fact, in the prospecting for hydrocarbons, shale is distinguished from sand and carbonate by the presence or absence of these elements, using Geiger counters in some cases. As discussed in chapter 1, hydrocarbons are most usually found in these other two mineral strata, not shale, and so identifying them is a key. Yet these concentrations are usually too small to be a health concern, and usually are below EPA threshold levels (King, 2012). The greater incidences, particularly of radium, appear to be in the Marcellus Shale of New York. Although usually low in concentration, if a crusty layer known as scale is formed anywhere in the system, these elements will have a tendency to concentrate in the scale.

Consequently, scale formation must be inhibited. This is why in the chapter encouraging the use of saline water in place of fresh water I suggest the removal of the ions of calcium, magnesium, and barium because these form adherent scale, especially barium.

If scale formation is inhibited, radioactive elements are not likely to be an issue (King, 2012). This is particularly the case if one reuses the flowback water, as I propose later in this chapter as a preferred option. Then any of these elements that come up go right back down where they came from.

Fracturing Chemicals

Rock is fractured by injecting a fluid, usually water, at very high pressures. This operation is made more efficient if the water contains a thickening agent to make it more viscous. This is usually a sugar derived from the guar gum seed, a crop that is largely sourced from India and Pakistan. To further improve the viscosity the molecules are bound together with a chemical known as a *cross-linker*.

The viscous liquid under high pressure creates fractures in the rock. Then, in order to remove the liquid, another chemical is introduced known as a breaker. It breaks the cross-linking bonds. Now the liquid is no longer viscous and can be extracted. A final step before the viscosity is dropped is to inject something known as proppant into the cracks. Proppant is usually sand particles but can also be synthetic ceramic materials. Its purpose is to hold the cracks open after the high-pressure fluid is removed. Proppant functions in a manner similar to the pillars and beams in coal mines, which enable the transport of coal-laden bins. As in the case of the mines, if the rock is not propped up in some way, the weight of the sediment above the zone will close the cracks.

Finally, after the fluid is removed, the gas will flow through the propped-open pathways into the main well bore and then up to the surface. Some gas will start flowing, mixed with the fracturing fluid as it is removed and the pressure drops. This mixture will need to be separated at the surface. This early gas, if not piped somewhere, is the primary source of fugitive methane into the atmosphere, as discussed in chapter 8, which deals with that topic.

Slickwater Operations

The vast majorities of shale gas operations do not use the thickening agents and associated chemicals. This is because the early production in the Barnett, in Texas, had dramatic falloff in production over time that was attributed to the plugging of the cracks by residues from the sugars used for thickening.

This use of essentially fresh water is known as "slickwater" production. Despite the imagery of slipperiness from the name, the fact is that plain water is not very slick at all and the friction in the pipes can be high. So, a friction reducer is added, which is usually a polyacrylamide, the same material that is used in baby diapers, wound dressings, and the like as an absorbing material. Also, simply greater volumes are needed in this case. On average about 4 million gallons of fresh water are used per well, but the use of 6 million gallons is not unusual.

Chemicals in Slickwater Fracturing Fluid

- **Friction reducer:** Always used. Usually polyacrylamide. Common other use: baby diapers, flocculent to remove fine particles in drinking water preparation.
- **Scale inhibitor:** Used about a quarter of the time depending upon solutes present. Usually phosphonate. Other common use: detergents.
- **Biocide:** Used in almost every instance. Glutaraldehyde, chlorine dioxide, mostly the former. Other uses: glutaraldehyde as medical disinfectant and chlorine in municipal water supplies. Will be replaced in part with surface treatment designed to kill the bacteria using ultraviolet radiation.
- **Surfactant:** Sometimes used. Many formulations. Other common uses: soaps, cleaners.

Other minor constituents of the fluid include scale inhibitors, for the reasons mentioned above, and biocides. This last is to address the fact that bacteria of certain types are harmful to the operation. Some species cause the formation of sulfur-bearing gases. In fact virtually all hydrogen sulfide found in reservoirs was formed by bacterial action. Also, certain other bacteria cause the formation of salts, which plug the pores and impede production. For all these reasons operators will not permit any bacteria in the fluid injected and additionally inject biocides, the principal one in use being glutaraldehyde. Other methods in some use include the use of ultraviolet radiation to kill the bacteria in the fluid prior to injection.

Use of Diesel and BTEX

One of the most vigorous arguments against shale gas production has been about the use of diesel in fracturing fluid. *There is no technical reason to use diesel in the fracturing fluid in shale gas operations.* When diesel was used in the past it was as a lubricant. Now polymers do the job well and are safer to handle. *The use of diesel for this purpose should not be permitted,* and such a rule will not impose any material hardship on operators. The class of organic compounds known by the acronym BTEX has also been cited as in use. These compounds may well be present in diesel, but if that is outlawed, BTEX constituents from that source should cease to be an issue and there is no technical need for them to be used in isolation.

Some reservoirs are water sensitive. In one type of sensitivity, the formation swells with water and impedes the flow of hydrocarbons. In this case, which is rare, propane or liquified petroleum gas (LPG; a mix of propane and butane) may be used. While diesel use is a possibility as the carrier fluid, there is no need for it since these other two work better at about the same cost. At a lecture I was asked by an environmental group official why industry did not use propane in place of water. My response was that to use an expensive hydrocarbon to produce another was something most would prefer to avoid. Keep in mind that propane pricing is pegged to oil, which is about four times the price of natural gas on an energy content basis. Transporting and handling propane for this use would open new concerns. There are some technical advantages to using liquid propane but they do not balance the cost and risk. An added element of cost is that the propane will blend with the produced natural gas and will need to be separated out. This requires cryogenic apparatus at the rig site. Because of these and other issues with alternates, water can be used safely as a fracturing fluid, and ought to be.

Diesel is also predominantly used as the fuel for the pressure pumps on the rig. The issues with this usage involve leakage, spills, and emissions. Pennsylvania has expressed concern because much of this activity is in populated areas. The emissions of concern in diesel usage, whether on a rig or on a city bus, are particulates. Metropolitan areas that switched public transport from diesel to compressed natural gas (CNG) have seen dramatically positive health effects. These are reported in chapter 15. In addition to the use of CNG, or liquefied natural gas (LNG) for that matter, another possibility is to substitute diesel in part or whole with dimethyl ether (DME) or even methanol, as discussed in chapter 16. Diesel engines today can substitute DME to 20 percent or more with no modifications. Engines can run on pure DME if modified.

While all this is known to some of the players, no attempt has been made to accomplish this, likely because of difficulties with infrastructure. Special effort would have to be made. States should take a good look at DME and CNG (or LNG) substitution and take steps to enable these changes. In the grand scheme of environmental hazard these are small potatoes. But you want to take the wins where you can get them. This one really has no technical hurdle, just the will to do it.

Disclosure of Chemicals Used

Put simply, *full disclosure must be made of all chemicals in fracturing fluid.* The only exception ought to be the use of a proprietary product, and for that exception to be given the *bar must be set very high.* Even in such a case the properties of the material relative to environmental issues must be fully documented. There are legal methods for confidential disclosure to federal and state authorities that protect all parties. If a patent is applied for, 18 months following the application the Patent Office will publish all the secret details anyway. The oilfield is famously slow to commercialize new products. An 18-month stay of disclosure is virtually in the noise. The proprietary product exception must not be a shield or an artifice.

The mandatory disclosure that is noted above is for the gross chemical, say polyacrylamide. No chemical is 100 percent pure. To identify each possible impurity would be onerous and may in fact not be easily performed. As long as the impurities have no material bearing on public health, their disclosure ought not to be mandatory.

In the run-of-the-mill shale gas operation, the chemicals used are well known, as described in the box on page 38. Nondisclosure by companies is likely on the advice of attorneys who fear the famous "slippery slope" concept. The way that goes is roughly analogous to the adage "give an inch and they'll take a mile." Advisors worry that if you disclose something now, even more will be asked for later and so on. Memo to attorneys who feel that way (and by the way not all do; this is not a diatribe on attorney conduct. Some of my best friends ....): without risk there is no profit. Second memo: when nondisclosure of something benign causes folks to get riled up, that is bad business.

One may well ask why allow such an exception at all. The answer is that we want innovation to design the most benign additives at the lowest cost. The reason diesel was used as a lubricant early on was that designers were not attuned to environmental concerns. The same goes for why fresh water is used as the base fluid rather than low-value saline water. We need to challenge the entire industry, and this goes beyond the oilfield, to design with an eye to sustainability: environmentally benign and using the least energy, while still being profitable.

If the industry is really clever about all this, it needs to be afforded the protection offered by law in the patenting process and to some degree in the trade secret maintenance process. Premature public disclosure kills patent validity. Patents will simply not be granted, especially abroad. We want the innovation and the associated profit to happen. In context we want it in the area of green chemicals and processes (Heintz & Pollin, 2011). The goal is to have functionally equivalent chemicals with minimal environmental impact. Norway has been using this approach for at least a decade: requiring green, yellow, and red labeling for oilfield products, and requiring a game plan (Norwegian Petroleum Directorate, 2010) to reach full green slates. One oil service provider has come up with a methodology for this for fracturing fluids. Another such provider has a fracturing fluid composition fully comprising additives from the food industry. Their CEO famously drank some quantity of the fracturing fluid (I assume the sand was absent) on a public occasion. He has since been seen in public. Incidentally, if my recommendations in the next chapter of using saline water are followed, he should take a smaller sip. Then again, who among us has not taken that inadvertent mouthful of sea water while body surfing? The astute majority who don't body surf is who.

Safely Handling Flowback Water

Some of the water and chemicals injected into the reservoir return as flowback water. Less than 10 percent of the chemicals injected return to the surface (King, 2012). The polymers degrade at reservoir temperatures, the biocides do their job and get consumed by bacteria, and other chemicals are partly trapped in the rock. Much of the water also stays behind. Recent theories suggest that water actually acts as a proppant in some fractures. In any case only between 16 percent and 35 percent of the water comes back to the surface. Although very little of the injected chemicals return, the flowback water now also carries with it some of the mineral content from the reservoir. The chief of these are chlorides of sodium, magnesium, and calcium, which account for the dramatic increase in salinity over what was injected. There may even be some aromatic compounds that are naturally found in association with some hydrocarbon deposits.

Flowback water is absolutely not suitable for surface discharge. It is also not suitable for being sent to municipal water treatment facilities, as is believed to have happened in Pennsylvania. These facilities are not equipped to handle the high salinity or the chemical content. Some have suggested the construction of special facilities to treat the water for discharge. However, the cost of doing that could be higher than that of two measures proposed below.

Deep Disposal

The least costly method of disposal of flowback water or any other wastewater is the deep injection well. This is an EPA-approved method known as a UIC (Underground Injection Control) Class II injection well. The key criteria are the porosity (it should be high) and permeability of the host rock, competent cap rock above it, and firm guidelines on casing and cementing of the well to protect freshwater aquifers. In most states these are monitored by the state environmental protection agency. The host rock is either a porous rock with no fluids or sometimes a rock depleted of fluid by prior production.

This type of disposal is very inexpensive in comparison to the value of the produced fluids. The fully loaded cost including amortization of the well is between $0.25 and $1.00 per barrel of wastewater disposed, with the former figure applying when the well is proximal and owned by the operator, and the latter including hauling some distance, and profit for the injection company. The water hauling is subject to spills since this is done primarily by trucks. A workshop held by the EPA in 2011 (US Environmental Protection Agency, 2011) has detail on this and other aspects of disposal.

In the last part of 2011 there was a lot of press about earthquakes in Oklahoma and Ohio presumed to be related to deep disposal wells. I cover those issues in greater detail in chapter 9. But there is no doubt that disposal wells should be planned and executed with care. Seismic studies must establish the absence of significant faults in proximity. There is evidence that some care is already being exercised; the suspect wells in Ohio were shut down promptly. Also, the EPA reports that there currently are 144,000 disposal wells in the US today, going back decades. So the occasional problem could have been expected, especially because the earthquake propensity was not recognized, and avoidance was not part of the design. Still, given that it is a regulated and monitored event, any such problem is correctible primarily through requiring that disposal wells not be in close proximity to active faults over a certain size. The specifications guiding this still need to be worked out.

Desalination Methods

The workhorse desalination method is reverse osmosis (RO). This uses a filter known as a semipermeable membrane: it allows only water molecules through and rejects others such as salt. Ordinarily the process of osmosis causes water to flow from the fresh to the salty side, essentially to equilibrate the concentrations. In reverse osmosis hydraulic pressure is applied on the salty side to force water to flow to the fresh side. Depending on the pressure applied, this will cease at some concentration of the brine, usually at about 75,000 ppm salt. At the end one is left with fresh water on one side and very heavy brine on the other. This is commercially used to produce fresh water from sea water in the Middle East, Australia, and other arid places. But the process leaves a brine to dispose of.

In coastal areas desalination operations put the resulting brine back in the ocean, although this practice is in review in some instances due to risk of damage to coral and other species. Certainly, in inland applications disposal will be a concern. Furthermore, this technique is fairly useless when the starting liquid is a heavy brine, as is often the case in shale gas operations.

Forward osmosis (FO) is a relatively new technique and has greater potential because it ought to use less energy. In this case a "draw" solution is designed with constituents that cause water molecules to be drawn to it from the flowback water side. Eventually the constituents causing this action are removed, sometimes by volatilizing them, leaving behind clean water. The flowback water side concentrates into heavy brine, which needs to be disposed of.

A relatively new technique is membrane distillation. This involves a membrane that transmits vapor only. Water vapor can be moved across the membrane, leaving behind the salts. In this case volatile organics, if any, would likely have to be removed first, else they too would go across. The allure of this technique is that low-grade heat can be used for the vapor production. One innovative method combines this with FO.

A popular method is to simply evaporate the water from the flowback water and condense it for use. This is energy-intensive because the latent heat of evaporation has to be provided. The resultant product is very pure. Some outfits are claiming techniques that minimize energy use. Something like this may be needed for the very heavy brines that are too salty for direct reuse and too salty for RO. But evaporation does get more energy-intensive as the salt content goes up.

Reuse of Flowback Water

This is the most elegant solution. But in order not to be cost-prohibitive, the industry has to tolerate greater salinity. This is discussed in chapter 7. Were we to require fresh water, the desalination costs would be higher. Flowback water salinities in the Marcellus range from 16,000 ppm to over 250,000 ppm. At the lower ends of that range no desalination would be needed, except possibly for the removal of the minor constituent divalent ions, for reasons mentioned in the aforementioned chapter. Even solids removal may not be necessary.

Surface discharge would require a good deal more treatment. So reuse in the fracturing operation would be cheaper than treating for surface discharge. Also, any minor impurities, such as radioactive species, which could be a problem for surface discharge, would present no hazard to the fracturing application. Bacteria would have to be removed because most operators are reluctant to inject these. But this too is being researched, and ultraviolet radiation for killing bacteria is already in practice (Warren, 2011). Alternative techniques such as the use of molecular iodine as the kill agent are also feasible (Chelme-Ayala, El-Din, Smith, Code, & Leonard, 2011). Iodine is at least 20 times as effective as chlorine and is more benign, being classified by the EPA as generally recognized as safe (GRAS).

In the Marcellus and Utica areas, very few strata have been qualified for deep injection. Consequently, this option is largely unavailable. Not surprisingly, this is the area, particularly in Pennsylvania, where the most incidents of irresponsible discharge or disposition have been reported. In recognition of this some operators are reusing the flowback water, proving that this is feasible. In many instances the cost will be very low, likely below the upper end of deep disposal cost. In any case, environmentally secure disposal is not optional.

If the costs are prohibitively high, researchers will innovate to get the cost down. Or only the more profitable reservoirs will be accessed until such a time as natural gas costs rise to profitable levels for dry gas production. As discussed in chapter 10, wet gas is extremely profitable and will easily sustain these costs of proper water handling. And there is plenty of wet gas in the Marcellus and Utica, the areas challenged by few deep disposal well options, and by far the largest and most prolific plays in North America.

Responsible disposition of flowback water should be mandatory. Until better alternatives are found and fully tested, the *only disposal methods must be UIC Class II deep injection or direct reuse.*

Zero Fresh Water Usage

"Drove my Chevy to the levee, but the levee was dry"

—From "American Pie" by Don McLean

Use of fresh water as the base fluid for fracturing operations is no longer necessary. Salty water with little or no value for human consumption or agriculture can be used. *Salt* water of convenience ought to be the default option in every case.

Shale gas wells use up to 6 million gallons of fresh water per well in fracturing operations. Only up to about a third of the injected water returns to the surface, so more fresh water needs to be added to make up the needed volume, even if the flowback water is reused.

A *Wall Street Journal* article (Gold & Campoy, 2011) tells a tale of water deprivation in south Texas. The town of Carrizo Springs, the imagery of plentiful water notwithstanding, is dealing with the conflicting demands of agriculture and gas development. Oil and gas leases often do not have water rights attached, although in Texas they do. Farmers without oil and gas leases have the option to sell their water to whomever they please. The value of the water is much greater to the oil companies, so they generally get the water they need. In the *Journal* report, data are presented for the number of water wells drilled for this purpose as opposed to all other uses. In some areas of Texas, 80 percent of new water wells drilled are in support of oil and gas operations. In total the number of wells in 2010 is about five times those drilled in 2005. The rate of this activity can only be expected to increase.

Heavy water withdrawals can strain aquifers, especially in drought years.

The major gas production areas of Barnett, Woodford, Eagle Ford, and Haynesville were in exceptional to severe drought in 2011. The prolific Marcellus in the Northeast is spared this particular problem. Nevertheless, water usage has been an issue in some communities, probably because water-related issues tend to have an emotional component. But in the other areas mentioned the issue is quite real. Curiously, the Texas Water Development

Board estimates that mining and oil and gas account for less than 2 percent of all water used. The majority is used for crops (56 percent) and municipalities (27 percent). But again, as the new game in town, shale gas drilling will inevitably be a focus of attention because the other uses already exist and are essentially unassailable.

The above numbers notwithstanding, the report of new water wells being dominantly devoted to oil and gas certainly indicates a growing trend. Fresh water drawdown can lead to the need to go ever deeper. At some point the water will get brackish, having salinities in excess of drinking water standards. This is because all groundwater tends to get saltier as wells get deeper. This is why it is rare to find freshwater wells deeper than 500 feet.

The exceptional droughts in portions of Texas are rare, this being the worst in several decades. However, in general a vertical swath through the country going north from west Texas through Arizona and Colorado and up to Wyoming is a perpetually water-deprived area. Important gas production areas other than those for shale gas are in Colorado and Wyoming. These are also in tight rock, albeit most commonly sandstone. Fracturing is required to economically release the fluids. This co-location of tight reservoirs with drought-prone environments is nature's little joke, as it were.

Salt Tolerance of Fracturing Fluids

The industry has been pursuing the objective of salt tolerance for a number of years. The initial driver was to improve the feasibility of reusing the water returning from the well after fracturing the rock. On the entire Eastern Seaboard of the US, deep disposal wells are not feasible due to the geology. Pennsylvania operations are rife with alleged instances of improper disposal leading to groundwater contamination. Absent a solution, production could well be halted. One solution is direct reuse at the site.

If only fresh water was acceptable, the desalination cost would be prohibitive in many instances. This is because the returning water could have salt content in excess of 200,000 ppm. The conventional desalination workhorse, reverse osmosis, is essentially useless because it needs to treat salinities in the vicinity of sea water (roughly 35,000 ppm) or lower, and the reverse osmosis *reject* water is at 80,000 ppm. Evaporative recovery, no matter how clever the design, faces the hurdle of supplying energy to overcome the latent heat of evaporation. So, the objective is to maximize tolerance to the salinity of fracturing fluids.

Fracturing fluids were originally designed to work with fresh water because such water was plentiful and it was easier to do that. Today salinities of up

How Fresh Does Water Have to Be?

Freshness is generally defined by the water's saltiness. The measure commonly used is total dissolved solids, or TDS. In the main these are chlorides of sodium, potassium, magnesium, and calcium. Drinking water is required to be under 500 parts per million (ppm). As a frame of reference, 10,000 ppm is 1 percent and sea water runs around 35,000 ppm, although that can vary from sea to sea.

Agricultural uses generally dictate salinity under 1,000 ppm. Curiously, though, the tolerance for chlorides is variable between plant species. At one extreme are date palms, which can handle up to 20,000 ppm, probably because of adaptive mutation to an environment wherein evaporation tends to render much surface water brackish. Sweet sorghum, a potentially important source of biofuel, is said to tolerate 3,000 ppm. Livestock, too, have variable tolerances. Sheep are the most tolerant, coming in at about 6,000 ppm for healthy growth. They can tolerate double that for a maintenance situation.

Salinity tolerances of livestock and poultry
In parts per million of total soluble salts

Animal	Maximum tolerance for healthy growth
Sheep	6,000
Beef cattle	4,000
Dairy cattle	3,000
Horses	4,000
Poultry	2,000

All of the foregoing underlines the fact that potable water is not needed for every application. In fact, the salinity should simply be fit for the intended purpose. This could be important in selecting the most suitable desalination technology.

Also, some saline aquifers could be moderately useful. In fact, one could seriously consider selecting farm products to suit the available water, rather than the conventional approach of treating water to be fresh. Edible plants can be genetically engineered to be more salt tolerant. One extreme is the class of plants known as halophytes, which actually preferentially consume salty water.

Similarly, commercial processes could be modified to accept higher TDS. One such is the fluid used for fracturing operations. More on that on page 46.

to 40,000 ppm are tolerable for fracturing fluid formulations, with some adjustments to the other chemicals used. There is little doubt that the *tolerance* can be raised to over double that figure. Some of the chloride ions impair the effectiveness of the cross-linkers and friction reducers. Of course cross-linkers are used only when sugars are used as thickeners, and this is largely not the case in shale gas production. Nevertheless, alternative chemicals can be employed in the presence of salinity, whenever cross-linkers are required. As mentioned in chapter 6, in the majority of instances the fracturing fluid is water with very few chemicals and no sugars. This is known as "slickwater" fracturing. The slipperiness evoked by the name notwithstanding, such fluids have high frictional losses. So, friction reducers are added. Some of these chemicals are less effective at higher salinities. However, replacement chemicals have been found.

Chlorides of sodium and potassium are particularly tractable. Those of magnesium, calcium, and barium are less desirable. Mostly this is because they will form an adherent scale in the flow system. Scale tends to concentrate radioactive elements if they are present. In some shale gas drilling areas, radioactive radium, thorium, and potassium are found in the formation and may be present in very low concentrations in the returning fluid. While these quantities are generally benign, if concentrated in scale they can present health hazards, especially during scale cleanup.

One solution is to simply remove these elements, known as divalent ions, from the fluid prior to use. The process for accomplishing this is very straightforward and is commonly known as water softening. Most municipal water systems and some homes with their own wells employ this process when the water is known to be "hard." In a domestic situation this is done largely because the divalent ions interfere with detergent action, so soap does not lather up effectively. At any rate, this is a known process and the only consideration is the cost of doing it.

Sources of Non-Fresh Water

To this point I've discussed the use of saline water in place of fresh. In point of fact a number of other sources of water are feasible for use. These include wastewater from industrial processes such as mining, cooling towers, and effluents in general. Samples of these would have to be evaluated and treated for use. A much more practical solution is the use of water from saline aquifers. As discussed earlier, all groundwater gets salty at increasing depths. So, one could expect saline bodies of water to be fairly ubiquitous. This is in fact the case, as exemplified by the map shown in Figure 6, which was plotted by the US Geological Survey about 60 years ago. The white spaces simply represent areas not investigated by them, and not necessarily lacking in such deposits. This is an important point because a lot of the white space is in areas covered by the Marcellus and Utica, both very prolific shale gas regions.

Figure 6. Depth to saline groundwater in the United States

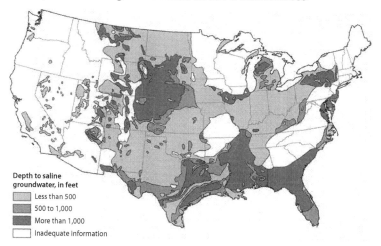

Source: US Geological Survey, 2003, generalized from Feth and others, 1965; Department of the Interior/USGS

Note that in many instances the water is shallower than 1,000 feet, making it very accessible. In general, shallower deposits are less salty. However, in these instances care has to be taken to understand the hydrology relative to adjacent freshwater deposits. In many cases, these are in communication, and withdrawals from one can affect the other.

Saline aquifers can be expected to be in reasonable proximity to producing areas. Consequently they represent reliable sources of water. The other alternatives, such as wastewater, mentioned earlier, while chemically acceptable, could vary in supply. Also, a given saline aquifer can be expected to deliver uniform quality over a long period of time, thus allowing the treatment processes to be standardized.

Assuming widespread acceptance of these ideas, the work needing to be done includes characterization of saline aquifers in producing zones and filling in of the white spaces on the map. The principal points of interest are concentrations of divalent ions and bacteria.

Bacteria will need to be eradicated before the water can be used as fracturing fluid. Bacterial species present in the shallow aquifers are likely to be different from those in the gas reservoir. Injection of unfamiliar species could have unintended consequences. We do know that certain bacteria are harmful to the reservoir from the standpoint of causing the formation of chemicals that plug the pores and impede fluid movement, and hence production. Other bacteria can cause the synthesis of hydrogen sulfide. Removal of bacterial species is straightforward. Reuse of flowback water will need some such process anyway. In general, one would expect those process steps to work hand in hand with the preparation of saline water for use.

Horn River Experience

Lest the foregoing appear to be a theoretical exercise, at least one outfit is using these methods routinely. This is Apache Corporation in their Horn River field in British Columbia, Canada. Incidentally, Horn River and Montney in British Columbia are on par with five of the largest fields in the US.

The driver for Apache was at least in part the difficulty of access to fresh water in winter. The water has to travel a considerable distance, and keeping it from freezing is a challenge. They access the Debolt saline aquifer, with salinity running in the vicinity of 35,000 ppm. The bacterial content is low and the mineralogy is acceptable. After minimal treatment, they are able to use it as the fracturing fluid. They sometimes blend it with produced water or even fresh water. They report that water from the saline aquifer provides a cost savings of over 50 percent over the use of fresh water (King, 2011).

The Apache experience is a concrete example of the use of brines of convenience. It would be reasonable to expect that a majority of shale gas operations could virtually eliminate the use of fresh water. From the standpoint of energy policy, this should be the new norm. Departures from this practice ought to require defensible arguments.

Is Natural Gas Indeed Worse for the Environment Than Coal?

"You may be wrong but you may be right"

—From "You May Be Right" by Billy Joel

Until recently, natural gas was seen as an indisputably cleaner alternative to coal. Robert Howarth and colleagues at Cornell changed all that, at first abortively when their study was demonstrably flawed. Their revised report, which now includes the contribution of fugitive methane in coal mining, was published in the peer-reviewed journal *Climatic Change* (Howarth, Santoro, & Ingraffea, 2011). Their thesis is that about 2 percent of the ultimate recovery of natural gas is released to the atmosphere. Since methane is many times more potent than carbon dioxide as a greenhouse gas, they compute that the net effect is worse than from the use of coal.

Hailstorms of criticism notwithstanding, some of the issues beg debate. A more recent study ("Switching from Coal," 2011) appears to be in support of Howarth et al.'s contention as well. In contrast is the report by the Worldwatch Institute ("Despite Methane Emissions," 2012), conducted in collaboration with Deutsche Bank, which unequivocally concludes the superiority of natural gas but nevertheless recommends attention to fugitive emissions. Howarth's own colleagues at Cornell University conducted a peer review and pronounced it flawed (Cathless, Brown, Hunter, & Taam, 2012). At last count only one supporting study, cited above, is counterbalanced by at least five in serious opposition.

A blog by Michael Levi (Levi, 2011) of the Council on Foreign Relations is worth reading mostly because Howarth surprisingly chose to respond to his critiques, and the back and forth is instructive. The most critical report is one by Mary Barcella et al. from CERA, entitled *Mismeasuring Methane: Estimating Greenhouse Gas Emissions from Upstream Natural Gas Development* (Barcella, Gross, & Rajan, 2011). This report alleges that key IHS data were "misused and severely distorted"—strong words. On balance all of the criticism centers on overcounting fugitive methane emissions from shale gas production operations.

So what is the public to make of all of this? They are right to assume that science is deterministic, at least in the broad swaths of the argument in question. When combusted, natural gas produces about 45 percent less carbon dioxide than does coal in producing the same amount of electricity. The Howarth study does not take the analysis to post combustion, thus intrinsically favoring coal because coal has lower combustion efficiency. The salience of such a comparison lies in the fact that cheap shale gas is being considered to replace coal in aging plants.

A fair comparison would be the best new gas plant to the most advanced coal plant because when the old coal plants are retired, the choice will be the best of the alternatives. A combined cycle natural gas plant has about 60 percent efficiency compared to about 43 percent for combined cycle supercritical combustion coal plants. This simply cannot be ignored and would skew the Howarth analysis toward gas even if their data were not suspect.

Howarth et al. defend their choice on the grounds that natural gas is used for purposes other than electricity generation (see commentary in the Michael Levi blog cited above). While this is true, coal is used predominantly for power, so in choosing to compare natural gas with coal, post-combustion analysis is appropriate.

Where the dueling reports diverge is in the area of fugitive emissions—releases of methane during the operations involved in producing and transporting the fuels. The quantities are in dispute, not the origins of the emissions. The Howarth data are from a variety of sources of variable quality. The most recent EPA estimates are much lower, as are those, predictably, from the gas industry.

Methane is about 25 times more potent than carbon dioxide in its global warming proclivity if the period under study is 100 years. The bulk of the debate surrounding the Howarth et al. work has been around the unconventional time scale of 20 years for the analysis, which most disadvantages methane. When carbon sequestration in deep saline aquifers is considered, the yardstick the practitioners are held to is well in excess of 100 years. In other words, the sequestered gas has to be guaranteed to not leak over that period.

In the case of coal, the emissions comprise methane found in association with the coal. For centuries this has been a known hazard of coal mining,

Global Warming Potential of Methane

Methane in the atmosphere will absorb infrared radiation attempting to escape to space. In so doing it causes atmospheric warming, much as does carbon dioxide, albeit at a lower level. Consequently it is classified as a greenhouse gas.

Over a 20-year timeframe, the global warming potential of methane is 72 times that of carbon dioxide. This means that if equal weights of the two gases are released into the atmosphere, methane will be 72 times as effective in warming. After about a 10- to 12-year period, the methane reacts with hydroxyl radicals in the atmosphere to produce carbon dioxide and water. This reaction product continues to warm the earth long after the methane has been consumed. Over a 100-year timeframe, methane has a net effect 25 times that of carbon dioxide, as compared to the factor of 72 in 20 years. The time period of comparison is therefore critical. Experts debate the appropriate timeframe, although current consensus leans to the higher number and is the number used by the Intergovernmental Panel on Climate Change (IPCC).

both from the standpoint of a poisonous atmosphere for miners and from the possibility of explosions in confined areas of the mines. In the past, canaries were famously used as indicators of methane. If they died, you got out in a hurry—a sort of go/no-go device. As previously mentioned, methane has no odor; commercial producers deliberately introduce an odor into methane for public use precisely for the detection of leaks.

Methane emissions from coal mining vary depending upon the nature of the mine. Deeper coal mines need to be purged for safety reasons, and therefore more methane is produced from them. The gas is purged with air, and then transported to the surface in vents. Consequently the methane is very dilute, usually less than 1 percent. This is the dominant source of methane from coal mines and is currently released to the atmosphere. Recently attention has been paid to capturing and utilizing the methane (Somers & Schultz, 2010). By many accounts much can be accomplished here economically. The next biggest source is when holes are drilled to purge the mines prior to actual operations with personnel. Mountaintop and other open pit mines likely have fewer emissions, but these are also harder to assess.

The Howarth study, flawed or not, has on balance been very good for all concerned. This is because it drew attention to the issue and forced a number of other studies that might not otherwise have been conducted. Such attention would likely not have been given were it not for the controversial comparisons with coal. With a large number of coal plants scheduled to be mothballed over the next few years, natural gas as a transitional fuel to more sustainable sources has great currency. Since that is the primary issue, Howarth's not taking into account the combustion step is a significant error. A direct consequence was the strident response, in the form of studies, from the likes of Worldwatch and the National Resources Defense Council.

The brouhaha certainly has focused attention on the fact that something concrete ought to be done about fugitive methane emissions. This ought to go beyond coal mining and oil and gas production. It should include the following two sources:

- According to the latest available EPA estimates, the third-largest US source of fugitive methane is livestock; in Europe and Canada it is the principal source. The rumen, or forestomach of animals classified as ruminants, converts feed by bacterial action known as enteric fermentation. One byproduct of this action is methane, which is expelled by the animal. The methane formation represents about a 7 percent loss of efficiency in food conversion. It seems the only truly viable approach to mitigating this is not capture, but amelioration. Modeling has demonstrated that up to 40 percent reduction may be possible by various interventions (Benchaar, Pomar, & Chiquette, 2001). A 2007 European study on dairy cows claims 27 percent to 37 percent reduction in methane through addition of just 6 percent lipids in the feed (Martin, Ferlay, Chilliard, & Doreau, 2007). For this study they used linseed lipids at a dairy farm. Since methane production represents inefficiency, cost-effective reduction is in the economic interests of the farmer.

- Methane from landfills is the second largest source in the US, and some estimate it to be the largest. In the absence of sufficient oxygen, organic matter decomposes to a mix of gases, about half of which will be methane. Intervention can be in two forms. One would be to divert organic matters such as household food scraps and yard waste to a specialized facility. Until such measures can be instituted, the alternative, capturing the methane from landfills, is feasible. This can be flared, which is burning at the end of a pipe, the resulting carbon dioxide being less harmful than methane, or it can be combusted for a purpose. This purpose could conceivably be generation of electricity, hot water, and the like. The business model of landfill operation does not necessarily fit this type of operation.

Natural gas production and distribution can cause the leakage of methane in two principal areas. One is in transportation. In some cases the high-bleed design of pneumatic systems in the pipeline infrastructure is the culprit. Replacement with low-bleed designs is feasible, and substantial gains can be achieved by doing this. A higher cost solution is replacing the gas with compressed air in the systems.

Also, the system of pipelines and associated valve assemblies can leak at various points after aging-induced malfunctions. But this can be addressed through maintenance mechanisms. One reason for the global drop in this sort of release is believed to be a Russian effort to replace old equipment.

The second and main source of fugitive emissions is the natural gas produced before a pipeline is in place to move it. This occurs in the early days of the prospect. Even in areas riddled with pipelines, a spur line to the new rig in question does not exist at the outset. Some operators may choose to not invest in a spur line until the reservoir is proven commercially viable. In that case, the initial gas produced during the discovery process has nowhere to go. It is often released.

A simple solution would be to flare it. This would dramatically reduce the problem since the released pollutant would be carbon dioxide, not methane. Howarth et al. assume that every bit of this gas on every well is vented. A recent study by Harrison (2012) of 1,578 shale gas wells across the country reports that in 93.5 percent of the cases, the spur line to the main gathering line is laid in time and the gas is recovered for sale. Thus only 6.5 percent of the gas in question is either vented or flared. This is a far cry from the 100 percent venting assumption of Howarth et al., so the situation is far less dire than they make out. The Harrison study was industry-sponsored, which could carry taint with some. But it is unlikely that the numbers are off by much. The larger operators such as Shell have publicly stated that they utilize all the gas.

The public may well ask why something useful is not done with the gas in every case other than simply putting it on the spur line, which, as we discussed is sometimes not present. The answer lies in part in the short duration of the production of natural gas wells. It cannot economically warrant any sort of capture and use. If such an economically enabling technology were to be developed, the potential to reduce methane emissions would be significant. The actual act of capture is straightforward and requires no particular innovation. The only issue is economical utility of captured gas as opposed to flaring. In some ways it is the same issue faced by landfill operators.

A little-known fact is that LNG is kept cold during the long voyage from Qatar or elsewhere by releasing small quantities periodically. This release causes chilling primarily through evaporative cooling. But in recognition of the economic value of the gas, it is collected and used in engines on the ship.

Natural gas associated with oil production when not in economically useful quantities is another potential source of emissions. Here, too, flaring is an option, as is any new technology to utilize the gas. Oil storage is another source, in that oil storage tanks are vented to release methane; in principle this methane could be recovered. In all recovery schemes, a yardstick for cost breakeven could well be the price set on carbon. Today there is no such thing in the US, but in Europe that price ranges from about $15 to $35 per tonne of carbon dioxide. One could use $30 per tonne as a target figure that one could reasonably project as an effective US-based "tax," no matter what the manner of implementation.

Just before this book went to press, the US EPA issued regulations (EPA, 2012) covering this issue. In summary, operators will be required to use measures to drastically reduce methane emissions during the early stages of production (the flowback period). They will have until 2015 to achieve this. In my opinion, this is a reasonable policy. To expedite execution, the industry would be well served by sharing best practices with the smaller operators. This technology is unlikely to have a proprietary component, so sharing ought to be straightforward. Although research is not involved, RPSEA (the Research Partnership to Secure Energy for America) could be instructed to execute this aspect. This would be a fit with the small producer assistance stricture under which RPSEA operates.

The new rules also govern some of the other areas we have discussed in this chapter, including oil storage tanks and valve systems. There are also specific reporting requirements, which ought not to be onerous and are needed to assure a concerned public.

A final note: Given that belching bovines are such a major part of the methane emissions equation, an outbreak of vegetarianism would help the environment!

Earthquakes:
Should We Be Concerned?

"Shake, shake, shake"
—From "(Shake, Shake, Shake) Shake Your Booty" by KC and the Sunshine Band
(written by H.W. Casey and R. Finch)

Fracturing operations produce seismic energy. These can cause moderate earthquakes if the operations are proximal to active faults. Conventional techniques can identify such faults and interaction with them can be avoided. With diligent pursuit there ought to be no concern.

In Lancashire, United Kingdom, an earthquake of magnitude 2.5 on the Richter scale was recorded on April 1, 2011. Despite the date being subject to some tomfoolery, this actually did happen. It was followed by a 1.5-magnitude tremor on May 27 of the same year. A study by European scientists concluded that it very likely was tied to fracturing operations. It also mentioned that this was due to a combination of circumstances unlikely to be repeated and that the maximum such that could be expected was a magnitude-3.0 tremor.

In 2008 and 2009 the town of Cleburne, Texas, experienced a series of tremors up to 3.3 on the Richter scale. This town had no recorded history of earthquakes, so the residents speculated that fracturing operations, which were a relatively recent activity, were the cause. A team of scientists from two major Texas universities concluded that fracturing was not the likely cause but that waste fluid disposal wells could be implicated, and the two were related in that the waste fluid was mostly from fracturing operations.

In late 2010 and early 2011 a series of earthquakes were recorded in Guy-Greenbrier area of central Arkansas. The US Geological Survey studied the phenomenon and dubbed it the "Guy earthquake swarm." Initial reports were not conclusive except to suggest that fracturing per se is an unlikely cause, but deep disposal wells could be the culprits. Of note, though, was the report that in Enola, southeast of Guy, swarms of about 3,000 quakes occurred in the early 1980s, and about 2,500 again in 2001, predating fracturing and deep disposal in both cases.

The Tohoku-oki earthquake off the eastern coast of Japan in 2011 was the fifth earthquake globally since 2004 of magnitude 8.5 or greater, an extraordinary recurrence rate considering that the last preceding earthquake of that magnitude was in 1965. The Tohoku-oki earthquake certainly cannot be compared with the others, in part because it could not possibly be human-induced except in a James Bond fantasy. But is the earth entering a period of increased seismicity?

The December 2011 issue of *Science* reports on a paper by Andrew Michael in *Geophysical Research Letters*: "Recent seismicity can be described by random and high variability of low-rate events within a Poisson process rather than a cluster of related events" (Mueller & Yeston, 2011). The main purpose of Michael's paper was modeling to show that the recent spate of large earthquakes, most lately the one mentioned above, was not an indication of an era of increased global seismicity. His point being taken, one could still conclude that higher incidences of small temblors were natural events and not anthropogenic.

But whether fracturing could develop enough energy to cause geological platelets to move deserves study. The same goes for stresses caused by deep disposal wells. But first some basics. Earthquakes are measured on the Richter scale, named after a Caltech professor who devised it. The information in the box comes from the USGS Earthquake Hazards Program website (http://earthquake.usgs.gov) and presents intensity in common terms. The Richter scale is logarithmic. This means that each increasing whole number indicates a 10× increase in displacement (ground motion) and about a 32× increase in energy. More recently, as shown in the box on the following page, the USGS has moved toward using the Modified Mercalli Intensity scale. Unlike the Richter scale, it has no mathematical basis. It simply is a ranking based upon the actual effects felt by the populace. USGS feels the public can better relate to such a scale. However, the popular press did not get the memo and still report these events on the old Richter scale. Some habits are hard to break. In an effort to be even-handed, I have provided both in the box.

The USGS table raises several points related to earthquakes observed in the vicinity of fracturing activities. The ones in the UK really ought not to even have been noticed. Perhaps an observatory recorded and reported it. The ones in Cleburne as well were barely in the noticeable range. But once noted, the recurrence would have raised concerns simply because they were not the norm. Not mentioned in the chronology above were some larger ones in Ohio, around 4.9. These appear definitely to be implicated with injection wells.

Earthquake Intensities and Frequencies

Magnitude	Typical Maximum Modified Mercalli Intensity	Annual Average Number
1.0– 3.0	I	1,300,000*
3.0–3.9	II–III	130,000*
4.0–4.9	IV–V	13,000*
5.0–5.9	VI–VII	1,319**
6.0–6.9	VII–IX	134**
7.0 and higher	VIII or higher	15**

* estimated
** based on observations
Abbreviated Modified Mercalli Intensity Scale:
I   Not felt except under very favorable conditions
II   Felt on upper floor of buildings
III Vibrations similar to passing truck
IV Felt indoors noticeably, outdoors sometimes; like heavy truck striking wall
V   Felt by nearly all, unstable objects topple
VI Felt by all, some frightened, damage slight
VII Poorly built buildings damaged, slight to moderate damage in well-designed normal buildings
Source: USGS, n.d.a and USGS, n.d.b

The other takeaway from the table is the sheer number of smaller events. The normal number worldwide in the intensity range of the suspect ones noted above is a reason to believe that many of those observed may be naturally occurring. Also, aside from a fear of the unknown, the moratorium on shale gas development called for in the UK based on those tiny quakes, the sort of which naturally occur a million times annually in the world, smacks of overreaction. This is especially so in a country reliant on imported gas. Single-issue activism without consideration of the other implications is always disappointing.

Seismic Activity Related to Fracturing

Many fracturing operations are closely monitored for seismic activity. This has been done for a number of years in order to map the pattern of fractures in an effort to improve drainage of the reservoir by making sure they knew where the fractures went. As a consequence, very detailed information is available. The magnitudes have always been known to be small, so the technique is designated "microseismic." In principle, monitoring wells could be placed in every new prospect to estimate the actual energy produced from fracturing in that rock. The most detailed studies to date are in the Barnett Shale. The levels observed are mostly under 3.0 in intensity.

The more important point is probably not the energy from the fracturing itself but rather the proximity to faults of reasonable size. According to the USGS, the magnitude of an earthquake is directly proportional to the length of the fault. 3-D seismic is a technique routinely used by the industry to delineate the subsurface. Essentially a three dimensional picture of the reservoir and the rock surrounding it can be produced. This is routinely performed in offshore tracts but not so commonly on land, where a more cost-effective two dimensional picture could do the trick. In prospective areas known to be significantly faulted, the prudent approach would be to run a 3-D survey to assist in placing the wells in the most productive areas while also avoiding faults of significant size. In areas with some risk, at least the initial fracturing operations may need to be accompanied by microseismic monitoring in real time. If a fault is intersected, the entry will be detected and the operation can be shut down.

Seismic Activity Related to Disposal Wells

Disposal wells have been more directly implicated in tremors. In Guy, Arkansas, four disposal wells were judged to be close to previously unknown faults. After two of these wells were closed, the incidence of earthquakes greater than 2.5 dropped significantly, and a moratorium was placed on wells in a 1,100-square-mile area over the fault system. Similar measures were taken in Ohio after additional quakes were linked to disposal wells.

There is a general belief that disposal wells, and these could include wells for carbon dioxide sequestration, have the potential to create seismic activity. While even this may not be of a damaging scale, it is disturbing for the public unused to such activity. Since disposal wells are regulated by the EPA, planning and monitoring ought to be straightforward. This in fact is likely why swift action to shut down was taken in Ohio: the wells were being monitored by the state Environmental Protection Agency. Careful planning and execution are critical for disposal wells.

These wells are used to dispose of all manner of liquid waste, not just that from drilling operations. Considering just the latter, one ought to give serious consideration to reuse of produced water being the primary option rather than the secondary.

Should We Be Worried?

In short, no. The seismic activity from the act of fracturing is small in intensity. It will by and large be below the threshold of human detection except in unusual situations. But care should be taken to not operate proximal to faults over a certain size. There ought to be rulemaking on this point. The technology of subsurface mapping to identify the location and size of faults is well known and currently used for practical reasons: operators do not want to lose valuable fluid down these faults. The technology of real-time monitoring of fracturing is well developed as well. Best practices ought to be developed with respect to the minimum number of wells in each prospect that ought to have such monitoring implemented. Good subsurface mapping combined with procedures to ensure no intersection of fractures with faults ought to be mandated.

Disposal wells do present a larger risk of detectable tremors, but again not likely large enough to cause damage. But the intersection of active faults by any such disposal wells ought to be avoided using very similar techniques to those mentioned above. This being an EPA-regulated activity, the execution of new measures ought to be straightforward. Strong consideration should be given to incentivize reuse of flowback water rather than deep disposal of it. Aside from avoiding the problem of possible tremors altogether, this would put less of a burden on fresh water withdrawals because at least a portion would be reused. In many, but not all, instances the cost may not even be greater.

PART III

Economics of Production and Use

Is Shale Gas Production Indeed a Giant Ponzi Scheme?

"You're just too good to be true"
—From "Can't Take My Eyes Off You" by Frankie Valli (written by Bob Crew and Bob Gaudio)

A *New York Times* top-of-the-fold front page piece, "Insiders Sound an Alarm Amid Natural Gas Rush" (Urbina, 2011), discusses the profitability of shale gas wells and is very bearish on the prospects. The author is careful to use the term "Ponzi scheme" in a statement attributed to someone else. But as anybody with modest discernment knows, such a reference made in the early stages of a piece is at the very least a tacit endorsement.

I acknowledge the principal points: some in the industry worry about profitability, especially given the low gas prices in the last year or two. I present here a case for rejecting the pessimistic premise. These are early days in the exploitation of a completely new type of reservoir. Continuous improvement, as in any industrial endeavor, can be expected. In the case of shale gas, the learning curve is likely to be steep. In part this is because of the sheer volume of activity. Each well can be drilled and produced in as few as 21 days, so the financial risk is low and the number of wells will be large. The setting is almost akin to a factory, the type of enterprise amenable to rapid learning curves.

Production from Shale Gas Wells Declines Rapidly

True. The decline is steep, with a drop of 60 percent to 80 percent in the first year. (Conventional reservoirs decline 25 percent to 40 percent in the first year.) After year two, there is a gradual asymptotic decline. The mechanism is still being debated, but premature closure of the fractures is a likely explanation. This could be due to insufficient penetration of proppant into the formation. Industry is working on materials and techniques to cause improved and more sustained flow. A Rice University–originated product sourced from nanomaterial is in early stages of commercialization. This and other such products are being designed to be lighter than conventional proppants, while still being strong enough. Being lighter they can be expected to float out further into the cracks.

Refracturing

In this technique, new fractures are initiated in existing well bores, often directly on top of the old ones. In the few cases where it has been attempted in the Barnett, the results have been dramatic. Production rates after refracturing have reached and exceeded the original starting production. And sometimes they decline at the same rate as before. This is indicative of the possibility that new rock pores are being accessed.

Current research at the University of Texas indicates that the optimal time to refracture is two to three years after initial production (Sharma, 2010). The University of Texas study will also examine other factors such as precise location relative to the old fractures. One hypothesis is that closure of the fractures in the zone produced induces a stressed region, which discourages new fractures from going to that area. In some observations, the new fractures are seen to bend away from the depleted zones.

Somewhat ironically, a shortcoming of the resource, poor permeability (a measure of the ability of fluids to flow in the rock), may be why refracturing works. Ordinarily, poor permeability means less flow, and hence less production. Fracturing improves that. But if the fracture paths are impaired, as explained above, the gas does not get fully drained from adjacent rock. However, it remains available for new fractures, and is for all practical purposes from new rock despite being proximal. From the standpoint of economics of the prospect, all that matters is that each operation cause enough production to ensure a rate of return. The fast declines are not highly material if this economic threshold is met. One final point: refracturing comes at a fraction of the cost of the original well because no new well bore is drilled. So the newer gas has a cost basis that could be a third or less of the initial gas. This does wonders for prospect economics.

Wet Gas

There is a passing allusion to wet gas in the *New York Times* piece, but it deserves serious attention because of its dramatic effect on profitability. Wet gas is defined as natural gas with a significant component of hydrocarbon species other than methane, known collectively as "natural gas liquids" (NGLs). The principal constituents are ethane, propane, butane, and even larger molecules broadly named "condensate." As a detail, although ethane is lumped in with NGLs, it is actually a gas at ambient temperature. This distinction is important in the method of separation from methane and is discussed in a separate chapter.

The economic significance of NGLs lies in the spread between natural gas and oil prices. Natural gas, on the basis of energy content, is currently priced at about a fourth of oil. Decades ago their prices were in parity. Natural gas liquids, the "wet" part of wet gas, are priced in relationship to the price of oil. Condensate is at or somewhat higher than the price of oil, and butane is definitely higher than oil because it is essentially a drop-in replacement for gasoline. Propane is at a discount to oil, as is ethane. Ethane is the least costly, at about half the price of oil. The actual price varies depending on location and availability. But all these are vast improvements over the price of methane which is typically one-third to one-fourth the price of oil.

A typical Marcellus wet gas is reported by one oil company as pricing out about 70 percent over dry gas. Range Resources reports that at a flat $4 per MMBTU gas price (incidentally the average for 2010 was around this figure), its internal rate of return would be 60 percent. That is way more profitable than many conventional gas prospects. But you and I can do our own calculations (see box on the following page)—or you can skip that and go to the punch line on returns.

Wet gas profitability is shown to be more than double that of dry gas. A downside would be drops in the price of oil, thus reducing the NGL value. In chapter 2, "The Oil Plateau and the Precipice Beyond," I describe models in support of the belief that oil prices will remain high except for the usual perturbations driven by external factors. Another downside is wetness at the lower end of the scale mentioned, below the average figure of 7 gallons of NGL per mcf of gas. Keep in mind, though, the heavy discounting we did in our calculations relative to oil. Even accounting for the costs to clean up the liquids and transport them, the value of NGLs should be closer to the price of oil than our conservative assumptions.

Wet Gas Economics

The wet portions of North American shale gas deposits average between 4 and 12 gallons of NGL per thousand cubic feet (mcf) natural gas. Typical Marcellus wells run about 1,500 mcf per day, and an average cluster of wells (pad) may have 15 wells, giving daily production of 22,500 mcf per day. Using a figure of 7 gallons NGL per mcf, that yields 157.5 gallons, or 3.75 barrels NGL per day.

Ethane tends to average 60 percent of the NGL. (The full implications of the lowest value NGL being so preponderant is discussed in chapter 12, "The Ethane Dilemma.") For simplicity I will count all the other liquids priced at a discount to $100 per barrel oil, at $70 a barrel. Support for this is the EIA-sourced pricing figure in chapter 12. That figure demonstrates the recent history of NGLs at roughly 80 percent of crude price and, separately, ethane at about 50 percent of crude price. In both the NGL and the ethane, I have been more conservative than that. Ethane I will price at half of that, at $35 a barrel. Ethane prices out at 0.6 x 3.75 x $35 = $78.75. The other NGLs collectively are worth: 0.4 x 3.75 x $70 = $105. Total value of the NGLs is the sum of those two: $183.75. (Note that I call ethane an NGL, following industry practice, but in the calculation I make a distinction because the EIA figure split out ethane.)

To estimate the effect of NGLs, it is simpler to reduce the above figure to that associated with one mcf gas. That would be $183.75/ 22.5 = $8.17. This calculation states that the NGL associated with natural gas could add $8 to the typical price of natural gas these days, which was $4 per mcf in 2011. To be immensely conservative, let's halve the NGL value by discounting severely. That still indicates double the profits over dry gas. The actual market value of such liquids can vary by area, which is why I have chosen to be so conservative. But the moral is that **the wet component dominates profitability**.

The Marcellus Shale, the largest and most prolific of the North American deposits, has a wet character on its western side. The as-yet not important producing states of West Virginia and Ohio are advantaged in this regard, as is western Pennsylvania. The Utica Shale is described in chapter 4. It is a newly discovered province that promises to be bigger and more productive than the Marcellus, but these are early days yet to put any certainty on the size. Its productivity, on the other hand, is more of a sure thing because the Utica is set deeper and so can be expected to have higher natural pressure to drive the fluids up. On average it has wetter character than the Marcellus, again on the western side. The same three states are advantaged by this.

How Things Will Play Out

Given the facts above, expect the wet gas prospects to be produced first. Over the next few years, the price of dry natural gas will rise because of demand. Massive switching from coal-fired electricity to gas will occur. This is because even without a price on carbon, the all-in cost of electricity from gas is less than that from coal at gas prices below $8 per MMBTU. In a recent publication (Rao, 2011), I present a model predicting gas prices to have a lid at about $8. This stability will contribute to switching from oil to gas. The switches will include methane propulsion of vehicles and gas-to-liquids-derived diesel and gasoline. Chemicals traditionally derived from oil will switch to being produced from natural gas or the associated NGLs. These chemicals include ethylene and all its derivatives and propylene. Many of these products are imported today.

Over time, all this oil replacement, plus electric vehicles, will make a significant dent in our $400 billion annual imported oil bill, and hence our balance of payments. Importantly, gas prices will be less subject to the whims of the weather because heating and cooling will be an ever-decreasing component of gas usage. Also, with shale gas production heavily distributed and deep inland, hurricane-related disruptions to gas supply, and associate price spikes, will be a minor factor. But 33 of the 42 US/Canadian ethylene crackers are on the Gulf Coast. So hurricane-related reduction in the supply of chemicals that are based on natural gas is still not out of the question.

Demand creation will enable a gradual return to dry gas production. Some of the earlier plays are profitable at $4 already. But a rise in the floor price will ensure the supply that will be available when the consumption trends described above mature.

And one day *The New York Times* will have a page one above-the-fold piece on how shale gas transformed the US economy. Then I will wake up.

Chemical Industry Prodigals Can Return

"Get back to where you once belonged"
—From "Get Back" by The Beatles (written by John Lennon and Paul McCartney)

Consistently cheap shale gas and associated liquids will transform the US chemical industry. Production that fled these shores in the face of high and uncertain natural gas prices will return.

Recent unrest in the Middle East caused a spike in the price of oil, with immediate impact on gasoline price. The price of natural gas remained stable over the same period. This story line has repeated often and underlines the principal difference in these two essential fuels. Oil is a world commodity, while gas is regional. Also they serve largely different segments of end use. Consequently, the fact that today gas is one-fourth the price of oil (in energy content) has little relevance in the main.

However, if industry believes that this differential will hold for a long time, technology-enabled switching will occur. In this chapter I predict a shale gas–enabled future of gas at low to moderate price for a long time. At the same time I subscribe to the view of an upcoming plateau in oil production, which will drive oil prices higher. These two trends taken together assure a high oil-to-gas price ratio. This will cause systematic switching where possible. Industries that went abroad due to pricing uncertainty will gradually return. This will be good news for US balance of payments and job creation.

Predictably cheap natural gas stimulates a number of different avenues of endeavor. To begin with, a number of chemical and metallurgical industries have a high component of their cost tied up in fuel. The ones most affected are those with natural gas as feedstock. Also impacted are those in which oil can be switched to gas. One such industry, the production of ethylene from ethane rather than from oil-derived naphtha, is discussed in chapter 12. That particular compelling value move had an unintended consequence. Propylene production was impacted.

The Propylene Story

One of the derivatives of propylene, polypropylene, is ubiquitous in our lives: roofing, carpets, bottles, and bendable plastics, to name a few. For years, when oil and gas pricing was in greater parity, propylene was a byproduct of ethylene production in oil refineries. It is also produced by tweaking the catalytic cracking process, at the cost of a smaller gasoline fraction. A refinery can change the mix essentially at will, presumably based on the relative profit potential.

But with a worsening oil:gas price ratio, ethylene production increasingly switched to an ethane feedstock. Unfortunately this process produces very little propylene as a byproduct. So, as reported recently in *The Economist* ("Plastics Prices," 2011), in the last two years the price of propylene has gone up 150 percent. Propane associated with shale gas production is easily converted to propylene by dehydrogenation, much as ethane is cracked to ethylene. But propane is a high-value commodity and is priced near oil, so this is a costly approach. A predictably low price for gas will allow for plants dedicated to producing propylene from gas. At least three companies, Lurgi, Total, and UOP, have the technology at an advanced state. Just as in the case of ethylene production, domestic locations for these plants are very likely.

Ethylene

This is the dominant feedstock for the plastics and fiber market. When natural gas prices started fluctuating in the early part of the last decade, much of the cracking capacity was shut down and the likes of Dow invested in plants in the Middle East and Asia, proximal to low-cost feedstock. The US became a net importer of ethylene derivatives. Now the US and Canada have predictably low prices compared to most of the consuming nations. All the major players have announced reopening of old plants and the construction of new ones. This is driven by two factors. One is the virtual certainty of high-volume ethane availability at low prices, and the other is the predicted demand in the coming years.

In a few years, the ethane supply from the shale gas related NGLs will exceed the cracker capacity. As discussed in chapter 12, the newly built crackers could be in the Marcellus area, either as massive plants or as distributed small-footprint reactors. To get a feeling for how many crackers would be supported, I've done a calculation (see box at right) using typical figures for NGL content of Marcellus gas. The size of crackers being discussed today by companies seriously considering building these is 1 million metric tons (tonnes) per year.

I conclude that such a unit would be served by about 420 wells. Pads with multiple wells will be increasingly common. So the number of pad locations serving each cracker will be somewhere around 20 to 30.

Marcellus Assumptions

Typical values are presented below for the liquid-rich segments of the Marcellus Shale.

- One thousand cubic feet (1 mcf) of Marcellus shale gas would produce approximately 7 gallons of NGLs (the range is 4 to 12 in wet deposits across the country), ~60 percent of which would be ethane.

- Each Marcellus well produces approximately 1.5 million cubic feet (1,500 mcf) per day of wet gas (Utica shale would be higher, maybe double, in part due to the higher pressure at greater depth).

- Each well pad would contain around 15 individual wells on average. Pad drilling occurs only after full development is planned. Initially individual wells may be drilled. The economic, environmental, and societal benefits of pad operations will make them routine in most instances.

- Each well pad could thus produce around 94,500 gallons per day, or 2,250 barrels per day (bpd), of ethane (7 gallons x 0.6 x 1,500 x 15 wells = 94,500 gallons).

- A new 1 million tonne per year ethylene cracker would need 63,000 bpd, or 1.3 million tonnes per year of ethane feed (Seddon, 2010). Thus, a single conventional cracker could require the ethane output of 28 well pads, or 420 individual wells.

- A 63,000 bpd natural gas processing plant would be expected to have a capital cost of ~$700 million, or ~$11,100 per bpd ethane, or ~$540 per annual tonne of ethane.

- Annual operating costs of such a facility would be on the order of $35 million (Seddon, 2010), or ~$1.50 per barrel, or ~$27 per tonne of ethane.

Crackers will take a few years to get into production. Assuming an immediate desire, we need to look at 2014 to 2015 targets. Some estimates put Marcellus production in 2014 to be between 5 and 10 billion barrels per day (bbd), of which a quarter may be wet. So, assuming a figure of 7.6 bbd, the wet gas will be 1.9 bbd, yielding 317,000 bpd of NGL, of which about 190,000 bpd will be ethane. That will cover three 1 million tonne per year ethylene crackers.

Financing for crackers always requires guarantees on long-term supply. If dry gas prices remain low, the proportion of shale gas exploited that is wet will remain high simply to assure more profitability. So, there would be a logical basis for security of supply. Furthermore there is an expectation that the low-cost driver will provide good margins for export of the ethylene or its downstream derivatives such as polyethylene, PVC, and polystyrene.

Nitrogen-Based Fertilizers

Modern agriculture relies dominantly on synthetic fertilizers. The most important one is ammonium nitrate, which also has an unfortunate use in explosives as well (think Oklahoma Federal Building bombing). Another is urea, much of which is used in the production of rice. As a major producer of crops, the US is a significant user of synthetic fertilizer. Much as in the case of oil, we use a quantity disproportionate to our population: 12 percent of the world's usage as against 5 percent of the world's population. The primary feed for this fertilizer is ammonia, which in turn is completely dependent on natural gas, which accounts for 90 percent of the cost. The high and erratic price of natural gas caused over half of the industry to flee to other parts of the world with low-cost gas. Trinidad and Tobago is the largest supplier by far, followed by Canada and Russia.

Cheap shale gas is luring this industry back. Given the importance of food to the nation, it would not be much of a stretch to suggest that fertilizer is a strategic commodity and that domestic production is a welcome change. Prior to the flight abroad, the US was a net exporter of fertilizer. This could happen again. It might also not be off base to suggest the possibility of reduced food prices due to a consistently low fertilizer price. Were this to happen, the irony would not be lost that the last time the nation discussed the food/fuel nexus, it was the anxiety occasioned by beef prices rising due to the diversion of corn to ethanol.

Who Will Do It and Where Will the Jobs Be?

The chemical industry is vertically integrated by and large, in one direction or the other. Vertical integration involves being the producer of the feedstock as well. So, the major oil and gas refiners are also producers of the oil and gas. There are exceptions, such as Marathon, which is not a major producer of the fluids but is a huge refiner. Since production, known as the upstream, and refining, referred to as the downstream, are such distinct competencies, the decision to vertically integrate is a business one.

Oil field service companies perform the bulk of the work in the extraction of the fluids from the earth. With one singular exception, no oil field service companies participate in the processing. Even that exception went away with the divestiture of KBR by Halliburton a few years ago. A separate set of companies designs and builds refineries, crackers, LNG plants, and all other manner of chemical processing plants. In other words, vertical integration is uncommon in all aspects of the oil and gas service sector.

Oil companies have tended to vertically integrate from the very beginning. The big ones take it all the way from upstream to downstream to products and retailing. This is why you see consumer automotive products such as lubricants made by the likes of Shell. They are also into retailing, although many gas stations are franchised and not owned by the major oil company. As a curious point, although seemingly integrated vertically, the gasoline at an Exxon station may not have come from an Exxon refinery and almost certainly not from an Exxon-owned upstream facility. This is for two reasons. One is that refineries, no matter who owns them, will take feed from wherever they can get it. Refineries are finely tuned to accept certain input compositions, so the precise nature of the feedstock is more important than the source. The second is that gasoline distribution is most effectively accomplished using blending sites that are agnostic regarding the source, as explained in the box below.

BP Gasoline May Not Be from a BP Refinery

Gasoline blenders and distributors are located strategically to serve various filling stations. The fuel may come from any sort of refinery and standardization has rendered this possible. The tanker trucks from the name brand stations drive up to the distributor and accept a product with a known octane rating. Blenders also put in the needed ethanol. Incidentally, the infamous 50 cent per gallon ethanol subsidy was being awarded to the blender, not the farmer making the corn. At this point the truck will blend in the special chemicals that each company believes imparts all the qualities advertized at the filling stations: better engine cleaning properties and so forth.

In one sense, therefore, the boycotting of BP stations was misdirected, although the message value might have been present. The station being boycotted was most likely privately owned by a guy who was about as far from the Macondo oil spill as was a shoe salesman from the potentially questionable labor practices of a shoe manufacturer. The gasoline in question may not have any BP provenance other than the blending additives.

Another type of vertical integration is backward from the consumer product. In this case, the giants who make containers, fabrics, and all manner of plastic goods integrate back to make the key ingredient.

In all the discussion regarding the return of the chemical industry, the question arises as to where these jobs will end up. Certainly some of the jobs will be in the same locations as plants mothballed by the shift abroad. In this category falls the announced reopening of ethylene crackers in the Gulf Coast by Dow Chemical. But the new plants could well go closer to the source of ethane, as discussed in chapter 12. Bayer, Shell, and Total have expressed interest in crackers in the Marcellus area.

The domestic fertilizer growth will follow similar lines. Much of the current capacity is in the Midwest and the Gulf, and one could expect expansions to current plants. New plants, especially if dedicated to export, could go near the East Coast. For a bulk commodity such as ammonium nitrate, which also has transportation safety issues, proximity to end-use states could drive new plants to be in the breadbasket.

When a single commodity is a dominant portion of the cost of a product, comings and goings of plant locations are more to be expected than would be for processes with multiple commodities contributing to cost. Predictably low natural gas prices ought to persuade prospective owners of plants of a highly predictable long-term future. Such predictability keeps discount rates down and is a magnet for investor dollars.

CHAPTER 12

The Ethane Dilemma

"Why don't you stay?"

—From "Stay" by Sugarland (written by Jennifer Nettles)

The abundance of shale gas and the as yet limited demand have kept the price of natural gas low in North America. The price per million BTU (MMBTU) hovered at or under $4 for most of 2011 and hit decadal lows in early 2012. As conjectured in chapter 10, this is driving the activity toward wet gas production. In the principal producing areas more than half of the NGLs in wet gas comprise ethane. This abundance of ethane is posing the industry with a dilemma with respect to the most appropriate manner of exploitation. A good problem to have, but it is a dilemma nevertheless.

Ethane is a molecule with one more carbon atom than methane has. This crucial additional carbon allows for easy conversion to ethylene, which is ethane with two hydrogen atoms removed. Ethylene in turn is the basic building block of a host of useful products (see box on the following page). But ethane really has no other use except for increasing the value of natural gas. Natural gas nominally has 1 MMBTU per thousand cubic feet (mcf). But some sources, especially coal bed methane, can be as lean as 800,000 BTU per mcf. In these cases, ethane, with higher energy content, can be used to raise the value to a million or a bit more.

The processing of ethane to ethylene is done in plants known as crackers. The ethane dilemma centers on the fact that the crackers are located on the Gulf Coast, over a thousand miles from the workhorse gas production areas of Marcellus and Utica. To compound the problem, the majority of the ethylene-consuming factories making consumer and industrial products are located, you guessed it, right near the Marcellus and Utica fields. So one scenario would have the industry transport the ethane down to the Gulf Coast, convert it to ethylene, and ship ethylene derivatives all the way back again. This chapter is devoted to discussion of this proposition and alternatives, together with the technical and economic hurdles of each option.

Making and Using Ethylene

Ethylene has the formula C_2H_4. It is synthesized by stripping two hydrogen atoms off ethane, which is C_2H_6. This process is known as *cracking*. Straightforward polymerization yields polyethylene, which is produced in two forms—low density, for food wrapping and the like, and high density, for bottles and other more robust applications. Perhaps less obvious is the product vinyl chloride, from which PVC is made. Other common materials sourced from ethylene include ethylene glycol, an antifreeze, and polyester resins for carpets and clothing. Styrene is a derivative, from which polystyrene resins are made for containers, and also butadiene elastomer for tires. Readily apparent is the fact that ethylene is the starting point for a number of commonplace products. Consequently, increased domestic production of this chemical can have a major impact on much of the chemical industry. High-value exports could be the result.

Prior to this windfall of ethane availability, ethylene was largely produced by cracking naphtha, a byproduct of oil refining. These crackers have the ability to make gasoline or ethylene, with the proportions driven by relative pricing. When ethylene is made from naphtha, a significant fraction of the output is propylene, another important building block for the chemical industry. When ethylene is made from ethane, this fraction is not available. So, abundant ethane and subsequent cracking to ethylene will create a shortfall in propylene. The consequences of this are discussed in chapter 11.

Given a choice between naphtha and ethane as starting feedstock, the latter will be picked every time in North America because of the cost differential. Ethane tends to track natural gas on price, which today in early 2012 is priced at about a fourth of oil on the basis of energy content. The most recent figures from the US Energy Information Administration (EIA) are shown in Figure 7. Note that the NGL composite price is close to that of oil and tracks closely with oil. Ethane pricing, although significantly lower than the other NGLs, is beginning to pick up relative to natural gas. This could be in response to the chemical industry's recognition of its relative cheapness compared to naphtha, which would be slightly over the price of oil. If natural gas prices stay low, as expected, wet gas will continue to be the target and ethane will be in abundance, thus keeping the prices attractive.

Figure 7. Comparative spot price movements in natural gas, crude oil, and natural gas liquids

Source: US Energy Information Administration, April 29, 2011

Impact of Low Ethane Price

Consistently low pricing combined with strong availability is changing the landscape of the North American chemical industry. In a 2011 report entitled "Shale Gas and New Petrochemicals Investment: Benefits for the Economy, Jobs, and US Manufacturing," the American Chemistry Council, a trade-supported organization, made some bold predictions. They believe that ethane production will increase by 25 percent. They estimate a $32.8 billion increase in chemical production as a result, and an associated $132.4 billion increase in US economic output. Total estimated jobs created are 395,000.

The US now is extremely well positioned competitively with the rest of the world. The unit cost of production of plastic resins is now on par with Canada and only slightly over the Middle East. It is well below all the other consuming nations, including China, Western Europe, and Japan, in ascending order of cost. A scant six years ago the US was on par with Western Europe. Significant investment that had fled to the then low-cost countries is returning, contributing to the economic figures estimated by the American Chemistry Council. Imports will increasingly be replaced by exports.

Since all of this is so new, policymakers must feel they are in something of a fog. LNG producers such as Cheniere are seeking permission for exports, when not so long ago they were building import terminals. In early January 2012, Congressman Ed Markey of Massachusetts expressed concern to Energy Secretary Steven Chu that approval of the Cheniere request would result in increased domestic gas prices. That particular concern may not have a lot of merit because of the sheer abundance and accessibility of the resource. But this does raise the question as to what is the most advantageous export: a commodity such as natural gas or a processed product? There can be little doubt that it ought to be the latter, if feasible. In chapters 14 and 16 I discuss the technical and economic viability of methane conversion to transport fuel. Ethane-derived resinous material, methane originated fertilizer, and methane-sourced liquid fuel are the type of high-value products most suited to export. The bulk of the economic value stays in this country.

The Ethane Cracking Options

North America, excluding Mexico, has 42 ethane crackers. Of these, 33 are on the Gulf Coast, 4 in Alberta, 2 in Ontario, 2 in the Midwest, and 1 small one in Kentucky. So, the Bakken Shale oil field could supply Alberta, while Eagle Ford in south Texas, a rich source of ethane, is proximal to Gulf Coast capacity. But the Marcellus and Utica are essentially stranded. These are also generally acknowledged as the largest of the deposits. This analysis concentrates on these two sources.

The various options are presented in Figure 8. I will key in on only the first, third, and fourth rows from the top. The most conventional option is to transport the wet gas to a central facility and chill it to the point where all the NGLs drop out sequentially. The ethane is the last to liquefy and can be separated and then transported via pipeline to a cracker. As mentioned earlier, the only viable capacity is on the Gulf of Mexico coast. Much of this capacity had been idled in the early part of this century following the gas price shocks, and had fled primarily to the Middle East.

Another option is to separate the ethane at or near the production location and then pipe it to a central cracker in the region. So, for example, a dedicated cracker in eastern Ohio or West Virginia or western Pennsylvania could serve all of those areas.

Figure 8. Processing technology options for shale gas ethane

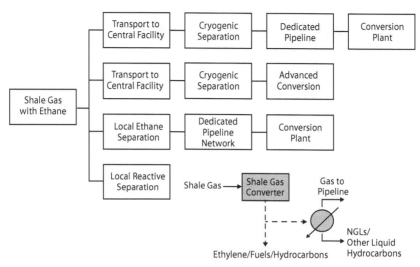

Source: Figure courtesy RTI International

The final option is the most innovative and also provides the highest returns. I refer to it as the local reactive separation option. This option involves reacting all the gases at the wellhead and tailoring the reaction to crack or convert the ethane while allowing the methane to pass through to the export pipeline.

The Gulf Coast Cracking Option

In late 2011, El Paso Energy dropped plans for an accelerated schedule pipeline that included using an existing portion. This was because the company was unable to get guaranteed contracts for gas. All major capital undertakings such as pipelines and LNG plants suffer from the need for these assurances of volume usage. But at about the same time, Chesapeake Energy, a major producer of gas, announced its decision to ship 75,000 barrels per day (ultimate capacity of 120,000) of ethane to Mont Belvieu, Texas, a distance of about 1,230 miles. West Virginia authorities responded with dismay.

"Every barrel of ethane that is shipped to the Gulf Coast means jobs and investment there and not here," West Virginia Department of Commerce Secretary Keith Burdette said. "We want to see those investments in this part of the country." But he acknowledged that absent an in-state option, producers have no option but to build a pipeline to a known market. The downstream oil and gas establishments clearly wish for a Gulf Coast cracker option. State governments will equally vehemently opt for the local option. In the end, the timeliest solution may be the winner. This may be the opportunity for the local reactive separation option (described below), provided it can be developed in timely fashion.

Overall capital costs for such conventional ethane recovery, transportation, and conversion would be on the order of $2,700 per annual tonne of ethane (assuming construction of a new 1 million tonne per year ethylene cracker on the US Gulf Coast). Overall annual operating costs for such conventional ethane recovery, transportation, and conversion would be on the order of $190 per tonne of ethane. Total annual costs of conventional ethylene production via shale gas extraction, ethane recovery, ethane pipelining to the US Gulf Coast, and ethane cracking would be expected to be on the order of $1,100 per tonne of ethylene (including capital recovery and shale gas extraction costs). This is in the right ballpark for current ethylene pricing, indicating such a conventional approach based on shale gas could be competitive with oil-based ethylene.

The New Proximal Cracker Option

This option is seriously being considered by Shell, among others. West Virginia authorities are keen to realize this option, and Pennsylvania and Ohio would like this as well, in their states of course. This option could reduce overall pipeline capital and operating costs by as much as 90 percent versus transporting ethane to the US Gulf Coast. Overall capital costs for this option would be on the order of $1,600 per annual tonne of ethane. Overall annual operating costs for this option would be on the order of $130 per tonne of ethane. This is decidedly cheaper than the Gulf Coast option, although the figure for the latter could be shaved with use of some existing facilities. More to the point, this option would require less pipeline cost and associated community pushbacks similar to the issues with the KeystoneXL heavy oil pipeline at the end of 2011.

But the most important reason for this option is job creation in general proximity to the production. Not to be ignored also is the fact that the highest density of polyethylene converters to useful products is in the Midwest. The Gulf Coast option would entail shipping the polyethylene back to near where the ethane came from. Most modern crackers are fully integrated to produce all the necessary downstream plastics. One could reasonably expect the proximal cracker to serve the polyethylene and other plastic needs of the East and Midwest. This option allows all the job and value creation to be proximal to gas production.

The Pad-Level Reactive Separation Conversion Option

A proposal under consideration is reactive separation conversion of ethane at the well pad level. It would need to have the following features. Based upon the assumptions shown in chapter 11 for Marcellus prospects, such a converter would need to be economic for throughputs of between 2,000 and 3,000 barrels of ethane processed per day. The output at a minimum must be ethylene, but more preferably a drop-in fuel for gasoline. That is a tall order for any conventional cracking process, but new technology appears available to achieve this, and RTI International believes it is feasible (R. Gupta, personal communication, March 13, 2012). Such a small-footprint convertor could be expected to be easier to finance and commission than a full-blown cracker. The full size would be on par with the pilot plant for a regular cracker, so time to market can be expected to be shorter than for most chemical processes. Timeliness with regard to jobs and the thwarting of raw ethane export ought to make it a favorite of state authorities.

The prominent producers in the Marcellus and Utica are not players in the downstream refining and processing area. So their imperative ought to be prompt monetization of the ethane. Currently, producing states are unhappy about plans to ship ethane to the Gulf but reluctant to antagonize a major employer. But given a viable local option, they may exert pressure and possibly provide investment incentives. To date the only local option available to them is a new dedicated cracker, and that may be in a neighboring state; also, the size and capital cost inevitably delay things.

The biggest hurdle, assuming techno-economic viability, is the business model. A large integrated cracker producing polyethylene is the current model. The independent polyethylene manufacturer is an endangered species. The reactive separation approach will require polyethylene to be manufactured in reasonable proximity. This is because the polymer is much cheaper to transport than is ethylene. The combined economics of all processes culminating in polyethylene will be the key, in comparison with the local cracker option. Of course, if the technical hurdle of the economic production of a drop-in fuel locally is crossed, that could trump all other options.

The solution to the ethane dilemma will most likely hinge on a combination of innovation and state activism.

The Alaska Pipeline Is Dead; Long Live the Alaska Pipeline

"The leader of the band is tired / And his eyes are growing old"

—From "The Leader of the Band" by Dan Fogelberg

Alaskan oil represents 11 percent of US consumption today, a vital component considering that the US imported over 50 percent of the need in 2011. The Trans Alaska Pipeline System (TAPS), a marvel of engineering built in 1977, is now in trouble. Another proposed pipeline, one to transport natural gas to the Lower 48, is on life support. This is the story of one that needs resuscitation and one that needs to be allowed to die. Curiously, these two outcomes are related.

A giant oilfield was discovered in 1968 in Alaska. But it was located up on the frozen North Slope. The sheer size and national priority dictated the construction of a pipeline (Figure 9). This was no mean task. The 800-mile traverse included three mountain ranges and hundreds of rivers and streams. Over half the length needed to be well above ground in order to not melt the permafrost, ground that is permanently frozen. The melting would occur because the oil flowing is relatively hot and needs to remain so in order to flow at a good rate. The above-ground pipe also needed to be built taking into consideration caribou migration and the like.

TAPS is the dark line from Prudhoe Bay (at top) to Valdez (bottom). The Chukchi Sea is on the upper left, and the Arctic National Wildlife Refuge (ANWR) is due east of Prudhoe Bay.

Figure 9. Map of Trans Alaska Pipeline System (TAPS)

TAPS is in trouble. At its peak it carried 2 million barrels of oil a day (MM bpd) and made the trip in three days, arriving at a temperature of about 100°F. Now, the falloff in production has reduced the flow to a third of that and it takes up to 15 days of travel. Figure 10 shows historical flows of crude oil in TAPS plus two projections: a continuous decline and a case where measures such as pipeline heating improve the flow.

The heat loss with the extended travel time causes the oil to be much cooler upon arrival. This increases the possibility of clogging and other problems. The practical limit at which it becomes economically prohibitive is debated but is somewhere in the range of 0.35 to 0.5 MM bpd, although the lower end of that range would need heat management solutions. That day is a scant few years away, 10 at most.

More oil is needed. The three candidates are heavy oil, oil from the source rock, and liquids converted from natural gas (GTL). Offshore oil from the Chukchi Sea and oil from ANWR (the Arctic National Wildlife Refuge) will not be considered here because politics, environmental hurdles, and location will render the output not timely for resolution of this problem.

Figure 10. Average daily volume of the Trans Alaska Pipeline System, in millions of barrels per day

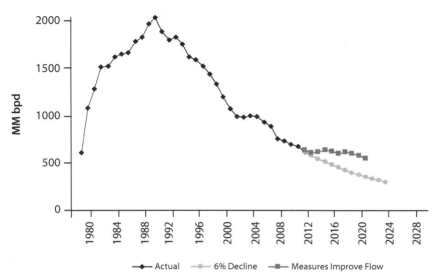

Source: Alyeska Pipeline Service Co., 2010

Heavy Oil

The Ugnu formation is estimated to hold about 20 billion barrels. The reservoirs are very close to the permafrost layer, so the oil is cold and hard to move. BP's Milne Point development has already demonstrated the ability to produce economically viable quantities using cold production methods. Such methods rarely yield more than 8 to 10 percent of the fluid in place. But even with this low recovery, the production will at least serve the purpose discussed above. Eventually BP will try some heating method because the layers adjacent the permafrost will be prohibitively cold. These methods necessarily will need to avoid damaging the permafrost. This is likely feasible.

The stuff is the consistency of molasses and not easy to transport. It would need blending with a lighter oil of some sort to be suitable for TAPS. The current oil from the Slope would do, as would condensate from the gas production. Another heavy oil source is the West Sak in the Kuparuk River field, also on the Slope. This has like quantities and is a bit lighter, more the consistency of honey. It too is close to existing infrastructure.

Heavy oil has a higher carbon-to-hydrogen ratio than does light oil. When refined, a significant portion of heavy oil drops out as a carbon residue known as petroleum coke, which has low value. Consequently, heavy oil sells at a discount. Heavy oils from Canada and Venezuela, major import sources to the US, also suffer from high sulfur and heavy metals. Alaska oil is much better in this regard. Nevertheless, only specialized refineries can take heavy oil. This will eventually prove to be a hurdle when the volumes down the TAPS are high. The good news is that California refineries, the closest to the Alaska port of Valdez, are generally well positioned to handle heavy oil refining.

Oil from Source Rock

The concept here is similar to our discussion with respect to shale gas. Proponents suggest that only about 20 percent of the oil "at source" found a suitable rock into which to migrate. They maintain that the remainder can be tapped much as is shale gas today, by fracturing the rock. In my discussion of the origins of shale gas, I, too, had conjectured that more would stay behind than would find a porous and permeable home. Now a little outfit out of nowhere, Great Bear Petroleum, has spent $8 million on substantial acreage. They will first target the Shublik formation, rock that is about 250 million years old and is the deepest of the three source rocks in the area. In part this target must be chosen because the deeper rock is more naturally pressured. Great Bear expects to access the shallower deposits in a later phase. This is not unlike my discussion with regard to developing the Utica first and then the overlying Marcellus.

The president of Great Bear has a BP (actually Sohio, back when he was there) Alaska pedigree as a geologist. So, this lends him some credibility. Great Bear expects to start production in 2013 and ramp up to 250,000 bpd by 2020. That would be ample to save TAPS. However, they are planning on year-round drilling, which is unusual up there. The logistics will be challenging, especially during spring thaw, with all those heavy trucks moving proppant.

There is every reason to believe that the source has the same mix as Eagle Ford and Utica: zones of oil, wet gas, and dry gas. One assumes Great Bear is chasing the oil leg first. But either their acreage or those adjacent must have the other two, which will have bearing on the following discussion. It is interesting that none of the current North Slope producers bid for the leases (except for a solitary lease by ConocoPhillips). The post–shale gas world knows all about source rock recovery. So, one school of thought would be that they will wait for success before bidding on adjacent leases yet to come up for sale. Or they plan to simply buy Great Bear, following a Lower 48 pattern in shale gas.

Finally, there is the ghost of Mukluk, which was predicted to be a giant reservoir. The company: Sohio. When the drillers got there, a billion (1983) dollars later, the oil appeared to have been spirited away. Dry as a fossilized bone. There is no geological parallel to the matter above. Just the memory of the last big one that got away. And this one is also big.

Shale Gas Kills the Gas Pipeline

There always has been natural gas in association with the oil in Alaska. With no ready market, it is pumped back in the ground. Estimates of gas reserves vary, but there are at least 35 trillion cubic feet in the vicinity of Prudhoe Bay. Another 100 trillion are estimated as potential for all the other areas, including the Beaufort and Chukchi seas and ANWR. For conversion to oil equivalent, one roughly divides by 6,000. So the total likely is about 22 billion barrels of oil equivalent. That is close to the light oil estimate for Alaska. The point is that the gas is a world-class resource and presently without a market.

The perceived solution has been a gas pipeline to the Lower 48. At least it was until shale gas came along. All of a sudden cheap shale gas rendered the Alaskan gas completely unnecessary. Fortunately for everybody, the bickering delayed the decision to this point when it is truly passé. According to the National Energy Technology Laboratory (National Energy Technology Laboratory [NETL], 2009), a gas pipeline would require daily throughput of 4.5 billion cubic feet per day for 35 years. That translates into a reserve of about 59 trillion cubic feet, which you will note is more than the readily accessible 35 trillion mentioned earlier. But in the end that is hardly the point. The Lower 48 is expected to be self-sufficient, and there is already concern regarding prices staying too low. Alaskan gas would not help that. But there is a remedy that leaves us with a TAPS that is saved rather than played on a trumpet solo.

The Gas to Liquids Option

A full discussion of gas to liquids in the context of cheap gas is in the following chapter, "Transport Liquids from Gas: Economical Now." But in short, it involves first reacting the methane with oxygen to produce a synthesis gas, a combination of carbon monoxide and hydrogen.

This reaction is straightforward, and the reaction products can also be obtained with coal as a starting point. The tricky part is then catalytically converting this mixture into long-chain hydrocarbons, also as described in chapter 14. During World War II, virtually the entire war effort was run on transport fuel using the Fischer-Tropsch (F-T) process. The Germans had domestic coal but not oil. Closer to the present day is the optimization of this process by the South Africans. During the apartheid-driven embargo, South African oil companies were forced to develop fuel with domestic sources. Today Sasol is arguably the leading purveyor of this technology, although the likes of Shell would certainly argue that point.

ExxonMobil, one of the big Prudhoe Bay property owners, has a process for taking synthesis gas to methanol and thence to gasoline, the so-called MTG process. But whether F-T is used or MTG, the prospect of a liquid fuel from Alaskan gas is real. The gas itself ought to be priced very low. This is particularly the case if it is gas associated with oil production. Using it means not incurring the cost of putting it back in the ground. One could even argue, tongue planted close to the cheek, that the price is negative.

The key question is how much gas capacity is needed to support a commercial scale GTL plant. A plant producing 100,000 barrels a day would be materially useful to keep TAPS open. The gas required for this is about 1 billion cubic feet per day. That is a far easier target than the 4.5 billion required for a gas pipeline, even were it to make economic sense, which it does not due to shale gas in the Lower 48. A comfortable target would be a 200,000-bpd GTL facility.

Things start to get interesting if one combines 200,000 bpd GTL output with a like quantity of heavy oil. In some ways this is near sacrilegious! GTL-produced liquids are incredibly clean; they are free of sulfur and all manner of impurities present in the same liquids from oil refining. But, getting past this squeamishness, the light fractions from a GTL plant would be very well suited to blending with heavy oil to make the latter transportable.

There are issues one would have to deal with. Sometimes when heavy oil is mixed with certain light hydrocarbon fractions, some of the carbon in the heavy oil will precipitate out as asphaltenes. You don't want that happening in the pipeline. But this sort of thing is well understood and simply needs to be handled. In fact, one option would be to use light molecules associated with the natural gas to deliberately precipitate the asphaltenes prior to shipping. A commercial process exists for doing this and is known as the ROSE process. In that case the GTL-derived product would simply be sent down the pipeline in batches between other oil batches.

The Trans Alaska Pipeline System is critical for US energy security. It must be kept open. There is currently a lot of rhetoric to the effect that this specter is being used by oil interests to open up environmentally sensitive areas. A future that combines production of heavy oil, blended with a liquid from natural gas, is one that we can all live with. This solution ought to keep the pipeline open for decades. If the shale oil option is realized, that would just be icing on the proverbial cake. That would merely raise the flow in TAPS back to the glory days' numbers. Alaska gas pipeline: rest in peace. A lone bugler can play "Taps."

Transport Liquids from Gas: Economical Now

"I can't get no satisfaction / 'Cause I try and I try and I try . . ."

—From "(I Can't Get No) Satisfaction" by The Rolling Stones
(written by Mick Jagger and Keith Richards)

Economical conversion of gas to liquids (GTL) has been repeatedly attempted by every major oil company over the years and always come up short of expectations. Shale gas now offers up a future with more certainty on gas pricing in North America, and possibly other parts of the world as well. Until now the only viable sources of gas for GTL were Qatar and Iran. The vast resource base, predictably low delivered cost, and distance from consumer gas markets made conversion of gas to liquids relatively attractive. All three of these conditions apply to Alaskan gas and the first two to shale gas in the Lower 48.

Expectations were high with the efforts of Shell and Sasol separately with cheap Qatari gas, but capital cost overruns dogged those projects, especially for Shell. The timing was unfortunate in that the majority of the construction was contracted during the prerecession boom and associated higher costs. There were also some reported teething problems on the process side. One assumes that at least these two companies have a handle on doing it more cost effectively the next time.

Industry kept pecking away with innovative catalytic steps, and today cheap shale gas may enable economical production in North America. Certainly abundant low-cost stranded gas in Alaska is fair game for the conversion, as discussed in the previous chapter.

In the late 1920s two Germans, Franz Fischer and Hans Tropsch, invented the process named after them and commonly abbreviated to F-T. A singular lack of oil in Germany combined with abundant coal led them to devise a means to obtain transport fuel from coal. This coal to liquids (CTL) process continued to be refined, but primarily for the gas to liquids (GTL) embodiment. The exception was Sasol in South Africa, which continued on the coal-derived path in response to the apartheid-induced embargo in 1987, which prevented oil imports. The Sasol work began well before that date in part due to the scarcity induced by the Arab Oil Embargo in 1973. As shown in the box, after the initial step of production of synthesis gas or syngas, the process steps are exactly the same for CTL and GTL.

Basic Principles of the Fischer-Tropsch Process

The starting point for these reactions is some carbonaceous material and a source of hydrogen. In the original discovery, coal was reacted with water in the following manner: $H_2O + C \rightarrow H_2 + CO$. This is sometime referred to as the water gas reaction. The reaction product is designated synthesis gas, or syngas for short. Syngas is a basic building block for a number of final uses, including simply combusting to produce electricity. In that case, often the reaction is taken further, as follows: $H_2O + CO \rightarrow H_2 + CO_2$. The hydrogen is combusted for power, and the CO_2 is disposed of in some way. If the CO_2 is sequestered, this process is the basis for clean electricity from coal, as in the federally funded FutureGen program.

The syngas may also be produced by oxidation of methane according to the following reaction: $H_2O + CH_4 \rightarrow CO + 3H_2$. This process is known as steam reforming.

In the F-T process, the syngas produced is tailored to a ratio of H_2 to CO depending on the catalyst used, usually close to 2. It is then reacted according to the following equation:

$$(2n+1)H_2 + nCO \rightarrow C_nH_{(2n+2)} + nH_2O, \text{ where } n \text{ is an integer.}$$

At $n = 1$, we get methane, which is certainly not desirable because that was our starting point in the case of natural gas as the feedstock. The key chemical challenge is creating the carbon-to-carbon bond, and this is more challenging for higher n's. The higher the n, the more energy-dense the liquid. Gasoline has n about 8, diesel around 12, and jet fuel in the range 10–15. All these fuels are mixtures with n as a range rather than a single number. The conditions of temperature and pressure determine the output. This reaction is very exothermic, that is, it generates a lot of heat, and this is a key process control variable.

F-T is the most versatile of the different methods for converting methane to liquids. This is because process conditions can be adjusted depending on the output desired. The starting feed for the gasification to syngas can be biomass as well, although typically preparation of the feed to be suitable for a gasifier would be a precursor step. Although not the subject of this essay, a new technology for converting biomass to transport fuel is worth noting. This is known as pyrolysis and involves treatment to a state short of actual combustion, resulting in the formation of a liquid. This liquid is very similar to crude oil and could be a direct feed to an oil refinery.

The MTG Process

One of the other technologies is the so-called methanol to gasoline (MTG) process conceived by Mobil about 40 years ago. This involves first going from syngas to methanol, which is a very simple process, and thence to dimethyl ether and finally to gasoline, which is a more complex catalytic process utilizing a zeolite designated ZSM-5. This reaction, too, is highly exothermic, so efficient removal of heat is a key.

The output is a gasoline ready to mix with conventionally produced gasoline. The MTG process was commercially operated in New Zealand by Methanex New Zealand Ltd. from about 1979 to 1996. The output was a gasoline with octane rating in the low 90s, in other words like the highest grades at our pumps today. The plant was then mothballed in 1996, with economics given as the reason. Since then ExxonMobil has improved the process and is offering it commercially today. The denser fuels (diesel and jet fuel) cannot be synthesized using this process. However, since gasoline is such a large part of the passenger car fleet, this is not an undue limitation.

Economics of Conversion

Consider a plant producing 65,000 bpd of output. For the statistics here I use the analysis of Michael Economides (Economides, 2005) for the input/output figures:

Input per day:
    Natural gas: 685 million cubic feet (MMcf)
Output per day:
    Diesel: 44,000 barrels
    Naphtha: 17,000 barrels
    LPG: 4,000 barrels

The selling price of the components can be highly variable. All refined fuels such as diesel have a value over that of crude oil which is known as the "crack spread." The term originates from the fact that large crude oil molecules are "cracked" to the smaller molecules that make up diesel or gasoline. The term n in the formula shown in the box is greater than about 22 for crude oil, and for gasoline it is around 8. The cost of this cracking plus a profit is termed the spread. This varies a lot over time and is one of the reasons that the price of gasoline at the pump sometimes appears not to be affected by the price changes of crude oil.

For diesel, a representative figure for the crack spread would be $15 to $20 per barrel. The spread for naphtha I'll estimate at about $5, and LPG I'll assume to be the same as light sweet crude. Based on these assumptions, and a crude oil price of $100, the value of the output is $7.24 million. The cost of the gas, at a price of $4 per mcf, is $2.74 million. Assuming a cost of $20 per barrel of output as the total cost of depreciation and operations, that adds $1.3 million to the cost, bringing the total cost to about $4.04 million, leaving a total profit of about $3.20 million. This is a very healthy margin, even assuming costs are higher in some places. But the main point is that it is sensitive to gas price. If the price goes to $12, as it did on occasion in the last decade, the cost jumps to $9.52, which is a severe loss position. Even $8 makes it marginal as a business.

This underlines the fact that shale gas–enabled consistently low prices would be a key to the business decision to build conventional GTL plants in North America. The other variable in play is the cost of sweet light crude. I assumed $100 in the analysis. Elsewhere in the book I support the models that predict oil prices as being consistently high and gas prices low. This is a recipe for widespread GTL. Certainly Alaska, with essentially stranded gas, is a prime candidate. Construction costs will be higher there, but the gas prices can be assumed to be a good deal lower than prices in the Lower 48.

Analyses by the National Energy Technology Laboratory have shown that liquids synthesized in Alaska could be sent down the existing oil pipeline in slugs between crude flows. I also suggest in chapter 13 that the transport of heavy viscous crude could be enabled by appropriate blending with GTL liquid.

Drop-In Fuels

This is a term that has come into vogue to indicate a synthesized fuel that can blend in with gasoline or diesel, or in some cases replace it. In some ways ethanol fits the description in that it blends with gasoline, but it has less energy density and so impairs a car's driving range. Butanol, on the other hand, would come close to being a true drop-in. A number of efforts are under way to produce such fuel from sugar ("The Future of Biofuels," 2010). I'll restrict my discussion here to fuels using methane as the feed.

F-T plants are required to be large to achieve economies of scale. The cost runs into billions. The Shell facility in Qatar produces 260,000 barrels of total fluid output per day and cost $18 billion to build. Future plants could be somewhat lower in cost but not dramatically so. This has two implications. One is that a large volume of gas has to be contracted for at least 20 years, and there has to be reasonable assurance of cost stability. The other is that the sheer size of the investment makes financing difficult. Consequently, industrial research has continually been searching for that elusive small-scale process that is cost-effective. One such breakthrough is being claimed by the company Siluria Technologies in San Francisco.

Oxidative Coupling of Methane

An approach that has been researched considerably is to attempt to directly couple two molecules of methane to produce a carbon-to-carbon bond of a larger molecule, ethylene. The desired reaction is:

$$2CH_4 \rightarrow C_2H_4 + 2H_2$$

This is highly exothermic, and temperature control is a key. The ethylene can be used for a number of applications, but further steps need to be taken to convert it to longer-chain molecules that can drop into gasoline. This is the goal of GTL.

This reaction has a low yield, and some component of ethane will also be produced. But the low yield is what has held back this approach. Now one company, Siluria, claims to have developed a unique catalyst to dramatically improve yields. Its website describes a process for creating a catalyst using an interesting amalgam of organic biochemistry and nano-catalyst dispersion in such a way as to yield a structure with highly reactive surfaces. If this indeed works as claimed, it will put oxidative coupling of methane into the commercial arena. One would then expect that modestly long-chain drop-in molecules could be produced for gasoline using much less energy than is needed for conventional GTL methods.

Nowhere else in this book have I gone into such technical detail, nor have I singled out any particular company. The departure is occasioned by the opportunity presented. The descriptor "game changer" is used with much abandon in the popular press, right up there with "paradigm shift." Here, though, is the prospect of a technology that could truly change the game. The promise here is of low-cost conversion with a small footprint. This last is the key. Even if the cost were to be the same as conventional GTL, the ability to conduct distributed processing at small scales is what is seductive.

This has two impacts. One is the ability to produce transport fuel all over the country, with minimal transport of the gas. The explosion of gas production will strain the pipeline systems, and utilizing at least a part of the gas locally will ease the problem. This will be particularly apropos in countries such as Poland and India, which don't have extensive pipelines. The second advantage is that the small scale of each plant eases financing and does not require massive gas supply contracts to be in place. As a bonus, these plants will be in gas-producing states, keeping the economic value local.

Natural Gas Vehicles: A Step in the Right Direction

"All in all it's just another brick in the wall"
—From "Another Brick in the Wall" by Pink Floyd (written by Roger Waters)

Direct use of natural gas as a fuel for vehicles could have a material impact on the reduction of imported oil. This is technically feasible and likely most applicable to commercial and fleet vehicles, especially for long-haul trucks. Light-duty passenger vehicle applications will need innovation to overcome some consumer-unfriendly features.

Some see natural gas as a transitional fuel toward a low-carbon future for North America. This can be manifest either simply in displacing coal for electricity production or in displacing imported oil for transportation purposes. The former reduces carbon dioxide emissions by virtue of the fact that the combustion of natural gas produces about half as much greenhouse gas as coal does. There is little argument on this point, except some argue that the act of producing the gas creates fugitive methane with adverse consequences; that is discussed in chapter 8. The more interesting avenue, substituting for oil in transport applications, creates debate. One approach would be to simply use the cleaner electricity to drive electric motors in cars.

Comparison to Electric Vehicles

The well/mine-to-wheel efficiency of an electric car is far in excess of any internal combustion engine, no matter what the fuel. So, simply in terms of using the least energy to drive a given number of miles, an electric car is the best option (see box on the following page for the calculation). This addresses a point that is not made often enough: we need to figure out how to get the same amount of gratification with less energy. Then where the energy comes from is less of an issue because associated emissions will be reduced, as a result of simply using less energy.

For example, today, due in large part to a California initiative commenced in 1971, television sets have standby power use of less than 1 watt. This is the power to simply keep the device in ready mode to allow the use of the remote monitor without a hard on/off at the machine. This little bit of couch potato convenience used to cost up to 12 watts. Similar standby wastage is in evidence in power drawn by devices left plugged in ready for use. This includes cell

---

Electric Cars Use Less Energy

To give support to this assertion, I will calculate here the efficiency of each step in the process and arrive at a fair comparison. I'll use the following facts and assumptions:

- A gallon of gasoline has 116,100 BTU, which equals 34 kWh.
- The average car being replaced delivers 35 miles per gallon (I am being generous here).
- For years the dogma has been that electric vehicles (EVs) use 0.2 kWh per mile. Nissan reports that the Leaf averages 0.25 kWh per mile. As in all electric and hybrid cars, stop-and-go driving gives better mileage than continuous operation. So, that number could be higher in some cases. I will use the 0.25 number for this exercise.
- Refining oil to produce gasoline consumes 20 percent of the energy in the oil.
- Coal-fired plants have efficiency of 40 percent (60 percent energy loss); by using coal, not gas, I am being conservative, and this figure is that of newer supercritical combustor.
- Electricity lost in transmission is 8 percent (a good estimate for the US).
- Energy to get the oil out of the ground is a wash with coal mining. Had I used the less conservative gas source for electricity, the offset would have been precisely correct.

So, energy losses for gasoline prior to its being consumed in the vehicle are 20 percent. Energy used after combustion is: 34 kWh in a gallon divided by 35 miles to the gallon, further divided by 0.8, equals **1.25 kWh per mile.**

Energy losses for EVs are 60 percent at the generating plant, minus 8 percent in transmission, equals 32 percent. Energy used by EVs equals 0.25/0.32 equals **0.78 kWh per mile.**

The ratio of the energy used to drive an average gasoline engine car to that used to drive an EV is 1.25 to 0.78, or 1.6. In other words, a conventional vehicle uses 60 percent more energy as an EV for the same purpose. Is this exactly right? Probably not, but it is not off by much. The key takeaway remains that the EV advantage has a facet that is not commonly recognized in quantitative terms.

phone chargers—but the worst offenders are printers. Overall about 8 percent of US power usage is for this bit of convenience.

But a fleet turnover to electric vehicles will take decades. In the meantime it would be well to also have alternatives capitalizing on cheap and abundant natural gas. One is processing it to produce liquids that drop right in as replacements for gasoline or diesel. This is viable and is known as gas to liquids, or GTL, discussed in chapter 14.

A potentially important automotive fuel derivative of natural gas is methanol. The most effective use will likely be in the form of M85 (85 percent methanol, balance gasoline). Once again natural gas must be seen as a bridging raw material. Unlike ethanol, production from biomass is very straightforward. But if shale gas remains cheap it may be hard to displace as the primary source for methanol. In turn, gasoline substitution will be very economical.

This chapter is devoted to a discussion of the pros and cons of using natural gas directly in existing or modified internal combustion engines. But one message is clear from the calculation in the box: other things such as cost and range anxiety being nearly equal, an electric vehicle is by far the preferred option from the standpoint of emissions. Not only is it more efficient, as shown, but the tailpipe emissions are zero. Sure, the electricity producer emits carbon dioxide, but capture at a plant is more tractable than on each vehicle. Having said that, the latter is not completely infeasible, and one attempt at doing so is being researched.

Natural gas for cars and buses is generally in the form of compressed natural gas (CNG). The gas is compressed to a pressure of about 3,000 pounds per square inch. This is about 200 times atmospheric pressure. The tank required for this has to be robust, which adds weight but makes it safe on impact. Also, research and development is ongoing to minimize weight and cost. A promising avenue is the use of adsorbents to store the gas at volumetric densities of a factor of 2 over CNG. The energy density is also lower than that of gasoline by about a factor of 4. Combined with the weight penalty, this causes range to be reduced. The only car designed to run only on CNG, the Honda Civic GX, has a range of something over 200 miles, depending on how you drive. That range is double that of the all-electric Nissan Leaf.

A purpose-designed vehicle such as the Honda Civic GX could well have the tank below the trunk, although Honda did not do that for the current model. However, retrofit vehicles will need to take up trunk space, and the space occupied is significant. Yet this has been done in taxis in New Delhi and

Kuala Lumpur. They simply install roof racks for luggage. The retrofit market is of interest for any quick uptake of this technology. The rate of uptake will determine the speed of installation of infrastructure.

Refuel Methods

Ease of refueling will be a key to acceptance. The uptake is much more likely to be swift for commercial vehicles than passenger. In many such cases, fleet refueling may be achieved simply on a dedicated basis. This has been the case for the cities in the world where all public transport has gone to CNG by fiat. Incidentally, those cities have documented health improvements, as noted on the following page.

There are two types of refuel systems: the fast-fill and the time-fill (an interesting euphemism for "slow"). Fast-fills are more expensive because they have an intermediate chamber to hold pressure between the compressor and the vehicle tank. Yet the time to fill is comparable to that of filling a gasoline tank. Time-fill systems take several hours, and the compressor is connected directly to the vehicle tank. This is the type used in home charging stations, which use domestically supplied natural gas as the feed. Honda partly owned a company named Phill that offered such a system. It recently went bankrupt, most likely because of slow uptake of the technology as a whole. It has resurfaced with different ownership. In the end, an overnight charge is not a prohibitive approach, particularly if commercial fast-fill stations are available for emergency situations.

Engine Type

The internal combustion engine for gasoline or diesel can accept CNG with very little modification. So, dual fuel supply is feasible and is in fact the case in most retrofits. But this means tolerating the combined weight and volume of a gasoline tank and a CNG tank. Also, conventional engines suffer about a 10 percent efficiency penalty using the gas. Dedicated engines can take advantage of the 125 octane rating of CNG. If the compression ratio is raised, more energy will be derived. The Honda Civic GX has a compression ratio of 12.5 to 1. This appears short of what is possible with the immense octane rating, but is likely a compromise on weight and cost. Also, some other combustion-related aspect might be limiting the compression ratio.

I have mentioned that fleet vehicles of all types are advantaged by CNG, as they would be with electric vehicles: charging/refueling infrastructure is simple. Public buses, delivery vans, postal vans, and so forth should find ready

---

Compression Ratio

This is essentially a measure of how much the fuel-air mixture is compressed prior to ignition of the fuel. At higher ratios, the pressure of the fuel-air mixture at the time of ignition is greater. This higher pressure results in more energy being delivered upon combustion, and the piston is driven harder. This translates into more energy to the transmission. However, if regular gasoline with octane rating 87 is used in a high-compression engine, ignition may occur prior to reaching full pressure. This is known as "knocking" because of the audible sound created. Loss of power and carbon deposits are the result. This is why higher compression engines need super gasoline, with octane rating of 93. Indy race cars have a compression ratio of about 17 to 1. They use pure ethanol, which has an octane rating of about 113, or methanol, with an octane rating of 117.

---

acceptance except for the retrofit time and cost. Fuel cost is low, and with shale gas the cost can be expected to be predictably low. Today reported figures are half the cost of gasoline per mile driven. This should be better with high-compression engines.

Long-haul transport cannot easily justify CNG because of its low energy density. Liquefied natural gas (LNG) is called for. This suffers a volumetric penalty of just 1.4, as opposed to 4.0 for CNG. However, refueling is much more problematic. Volvo is field-testing a truck using a 75/25 blend of LNG and diesel. Volvo claims the range to be between 500 km and 1,000 km depending on conditions. For European long-haul, that seems to be a comfortable distance to refuel. But again, LNG transport and dispensing is specialized. LNG for trains to replace diesel is also being piloted in India and elsewhere. Range is again an issue, and it will remain so until refueling infrastructure is more widespread. The key driver in India appears to be the cost of diesel rather than environmental gains.

Health Benefits of Substituting Diesel with CNG

The principal issue with diesel emissions is particulates. Engines and fuel have improved to the point that the old thick black exhaust is not really a factor any more. But the particulate loading is still high. An Italian study considered two scenarios in the period 2009 to 2011: a trend scenario of 3.3 percent replacement with CNG vehicles and an aggressive scenario with 10 percent replacement. In metropolitan areas they estimated a 1.3 percent reduction in deaths related to respiratory and heart diseases with the trend scenario and a

4 percent reduction in deaths under the more aggressive replacement scenario. Expected improvements included 4,197 and 13,115, respectively, fewer asthma attacks in adults. Similar figures were reported for lost work days and children's illnesses.

Such studies have also been conducted in New Delhi and Kuala Lumpur, in those cases with actual retrospective data. A New Delhi study (Akbar, Lvovsky, Kojima, & World Bank, 2005) claims 3,629 fewer deaths due to the switch to CNG by all public transport, which was ordered by the Supreme Court of India. This success has allowed expansion of the program to 22 major cities. Again, fleet operations are easier for implementation.

Europe is much more positively affected by the diesel switch to CNG than is North America, where diesel passenger vehicles are simply not in demand. That could change soon. Automobile manufacturers are realizing that the miles per gallon standard may most readily be achievable through partial fleet conversion to diesel with its 35 percent better mileage over gasoline. This may be the time for them to take stock of the CNG substitution angle. Unless the CNG tank is tucked away below the trunk, the US buying public is very unlikely to buy the vehicles. But light-duty trucks should be fair game in particular because there would be more design scope to include the CNG without intruding on the functionality.

Mandatory CNG for public vehicles (including taxis) and utility vehicles such as delivery vans in metropolitan areas, LNG for long-haul trucks, and possibly CNG for light-duty trucks: these are all worthy targets that could make a serious dent in our imported oil bill. Ultimately electric vehicles are the future, provided needed advances in battery cost and performance are made.

Advantage Methanol

"Mere alcohol doesn't thrill me at all. /
So tell me why should it be true"

—From "I Get a Kick Out of You" by Ethel Merman (written by Cole Porter)

As I have discussed in previous chapters, a cheap natural gas future is conducive to the production of liquids for internal combustion engines. Fischer-Tropsch synthesis gives us drop-in replacement for oil-derived gasoline or diesel. The methanol to gasoline (MTG) process likewise produces moderate octane gasoline, with methanol and dimethyl ether (DME) as intermediate products. Methanol can be processed to DME, which can be a cleaner burning substitute for diesel. But what of methanol itself? Could the supporting actor become the marquee attraction?

A careful examination shows that if certain measures are taken, methanol could be a viable endpoint. In fact, of all the gasoline substitutes derived from natural gas, methanol may well have the inside track based solely on economics. In tennis terminology, methanol has moved from the deuce to the advantage court. That is one step from winning the game. Harnessing the high octane rating would give it set and match against the perennial champion gasoline.

Widespread acceptance of any gasoline or diesel substitute requires three conditions to be met. The first is low cost per mile driven relative to the incumbent fuel, preferably without the assistance of subsidies. In the event of subsidies, the consumer would need some assurance that the post-subsidy price would still be attractive. In fact this assurance is likely more needed by the retailer who would need the security of long-term demand to incur the expense of a dispensing system. This brings us to the second condition, the need for a low-cost distribution infrastructure, including pumps at retailing stations within reasonable driving distances for most consumers. This last was one factor in the unpopularity of E85 (Kindy & Keating, 2008). The final condition required is that of easy availability of vehicles that are able to operate with the alternative fuel. In the limit, all new vehicles ought to be able to run on any mix of gasoline and alcohol alternatives. Bills in both chambers of Congress await passage that would ensure that most new vehicles after a period of time would be flex-fuel vehicles (FFVs).

Price of Methanol Compared to Gasoline

Today, in early 2012, methanol wholesale price is $1.13 per gallon. Distribution and markup will add 10 cents, and another 20 cents for federal and state taxes brings the total to $1.43. To compare with gasoline, this needs to be doubled because methanol has about half the calories of gasoline. Later I will discuss how this calorific disadvantage can be ameliorated or eliminated. But in the normal situation for use in a FFV, the effective cost of methanol is $2.86. On the day of this estimate, regular gasoline was selling for $3.79 as a national average. So, clearly methanol enjoys a cost advantage. Add to that another fact: methanol has an octane rating of 117. So the comparison ought to be with high-octane, or super, gasoline, not regular. In most states one would need to add another 25 cents, bringing the gasoline total to about $4.04. As a practical matter the most likely fuel used would be M85, comprising 85 percent methanol. But the substitution comparison based on the neat liquid is still valid.

All this does beg the question: Will methanol prices in the future enjoy this advantage? Since the bulk of methanol today is produced from natural gas, the question shifts to the forecast in natural gas prices. In fact, at a recent methanol policy forum, David Sandalow, Assistant Secretary of Energy for Policy and International Affairs, in touting the value of methanol, cautioned, "If natural gas prices increase substantially and price differentials between methanol and gasoline revert back to historical norms, the economics could be difficult." This is precisely the concern addressed in the analysis below. I conclude that methanol will be cheaper than gasoline on a per-mile basis for decades. The only wild card is extraordinary demand for natural gas outstripping the production capability.

The cost to produce methanol is plotted in Figure 11 as a function of natural gas price.

In March 2012 prices were essentially at decadal lows, driven by an abnormally warm winter. These prices are not normal; the October 2010 figure shown is more what one could have expected. That figure, incidentally, was roughly the average for 2010 as well. One readily observes that at those values, methanol can be produced for about 50 cents per US gallon. Keeping in mind

Figure 11. Methanol cost as a function of natural gas price

this is cost to produce and adding a profit of 15 percent gives about 58 cents. Add to that the aforementioned retailing and tax components and you get about 88 cents per gallon.

In chapter 3, I presented a model indicating a ceiling for gas price between $6.50 and $8.00 per MMBTU. Support for this is available from the work of Amy Jaffe and colleagues (Medlock, Jaffe, & Hartley, 2011) who used different scenarios of shale gas development. In none of their scenarios do they exceed decadal averages of $6.50 for the next three decades.

My forecasted ceiling is plotted in Figure 11 and represents the upper end of what one could expect for the methanol cost of production. At the highest end, methanol rings up at a pump cost of about $1.28 per gallon (85 cents to produce plus 15 percent profit plus 30 cents in distribution and taxes). An important point of note is that these calculations notwithstanding, the price of methanol will be driven by market factors. However, as one can see there is a fair bit of headroom between gasoline and methanol, and even accounting for variability in both, methanol is still a viable choice. Gasoline price will always be driven by world events, whereas methanol will be largely regional. In that sense one could expect more stability. Natural gas represents about 75 percent of the cost of methanol production, so stability in that commodity will at least keep cost under control.

Another factor is that methanol can be produced from coal and biomass, so alternative feedstocks are a moderating element. In particular, low-grade coal such as lignite is a useful feedstock. Lignite, available at the mine mouth for $25/tonne (recent figure from the EIA) will yield methanol at a cost of around 70 cents per gallon. So, if my forecast is vitiated by unusual demand for natural gas, plentiful low-grade coal can kick in. This coal is not particularly useful for electricity production because of its high moisture and ash. In fact, the high proportion of this useless component adding to freight cost is the reason I advocate mine mouth processing. With vast deposits in Texas and states northeast from there, the mine mouth is not far from infrastructure. Of note regarding the switch to another raw material is that the process after the formation of syngas is the same regardless of the feed character.

Methanol Production Cost

For this computation I used a standard plant producing 5,000 tonnes per day of methanol. This translates to about 40,000 barrels per day. As a frame of reference, the announced Sasol GTL plant for Louisiana is rated at a bit over double that figure. In general methanol plants require less capital than F-T process plants. The 5,000-tonne-per-day plants could be expected to cost about $800 million, whereas an F-T plant of the same capacity would be about two and a half times that. The time to construct would be similar in ratio.

This plant would consume 150 MMcf per day of natural gas. This is the output from about 100 Marcellus wells, or about 4 to 7 pads. Such plants could be distributed in the producing areas, keeping the jobs local. In fact, the simplicity of the process could well allow even smaller plants without much loss in economies of scale.

Viability of Flexible-Fuel Vehicles

The intent of an FFV would be to use any mix of gasoline, ethanol, or methanol. While there is some difference of opinion, most believe that the current FFVs could use methanol. Possibly a somewhat more robust fluorinated elastomer would be needed. Certainly the software or firmware would have to be modified to allow for a methanol mix. With an M85-filled tank, the range loss would be about 42 percent. In other words, a car with a range of 350 miles would now go about 200 miles with a full tank of fuel. The consumer would have to decide whether a lower price and lower emissions are worth filling up more frequently.

Today that lower price would be calculated as follows: 15 percent at an average gasoline price as noted above of $3.79 plus 85 percent at $1.76 (natural gas at $4 delivers methanol at this price after the doubling to take into account the fact gasoline has twice the energy content) equals $2.06 for a gallon of M85. I used the price of regular gasoline in the calculation. But M85 performance would need to be compared to high-octane super, priced at $4. Still, the consumer would pay half for the fuel in exchange for filling up almost twice as often. Eventually, if auto makers make larger tanks to accommodate this, then the fill up will be at a normal frequency. Most of the public would take that tradeoff; these are folks who are inclined to drive a mile or more for 10-cents-cheaper fuel. The comparison to the price of super gasoline is not completely fair because regular-compression engines get no benefit from the higher octane rating. The clever nomenclature "super" unintentionally causes some of the public to erroneously believe it is better.

The distribution argument goes as follows. If a large proportion of cars were to be FFVs, and if methanol had enough allure to get some fraction of these to owners to use it, a dedicated M85 pump would pay for itself. Furthermore, all the other cars on the road have either regular compression, needing 87 octane regular gasoline, or sporty cars with higher compression. These latter engines need 91 octane fuels. Cars currently in service do not need a third grade of gasoline. In fact the practice today of having octane ratings of 87, 89, and 93 makes no sense. Future pumps would be 87 and 91 octane gasoline and M85.

High-Compression FFVs

This is the future: smaller, more powerful vehicles with a longer range and, in the case of methanol, nearly half the carbon emissions for the same miles driven as with gasoline. The "Flex-Fuel Fairy Tale" (chapter 22) lightheartedly alludes to such things. But fantastic sounding though it may be, there is firm scientific basis for these assertions. All of it relies primarily on one feature: all three gasoline alternatives—methane, methanol, and ethanol—have in common the feature of extremely high octane ratings.

Our goal: an FFV that accepts any alcohol combination with gasoline, and also methane if practical. Of course the gasoline portion would need to be small or zero for the octane number to be high enough. Certainly E85, M85, and CNG would function in such an engine. But until there is a breakthrough in ethanol production cost and a similar advance in storing CNG in a smaller volume than currently, methanol will be the game. So, any further discussion will be for this fuel. But an FFV accepting both alcohols, and possibly methane, has the virtue of providing consumer choice.

How Gasoline Is Affected

Gasoline appears to be a loser on this, and some may argue that it would be premature to force a wholesale switch. As I discuss below, there is a variant that allows a mixture with a majority of gasoline in an effectively high-compression engine, known as the direct injection engine. Also, existing vehicles will be on the road for a long time. But if gasoline is forced into the position of being just another fuel, the future sought by Gal Luft and Anne Korin in *Turning Oil Into Salt* (Luft & Korin, 2009) will be realized.

Their thesis is that salt used to be a strategic commodity because of its critical function in preserving food. Wars were fought over it, and people were paid in it. The word *salary* comes from salt. Then refrigeration changed all of that. Salt became a useful, even essential, but not strategic commodity. Luft and Korin suggest that cars allowing fuel choice will render gasoline, and by extension oil, a nonstrategic commodity.

First let's discuss the direct-injected, alcohol-boosted engine. This is a relatively high-compression engine primarily powered by gasoline. When the likelihood of premature ignition is detected, methanol is injected using a special line and port. It has a high latent heat of evaporation, so when it is injected, it cools the chamber. This suppresses premature ignition, eliminating the knocking and realizing the high efficiency of the high compression with very little total methanol. Such a gasoline engine can give efficiencies equal to or greater than a turbocharged diesel, which is more expensive.

The greater promise from the standpoint of displacing gasoline is an engine with a much higher compression ratio, closer to 17. Any alcohol blend with low gasoline content would operate in such an engine. Work at the EPA laboratories in Detroit (Brusstar, Stuhldreher, Swain, & Pidgeon, 2002) and at MIT (Bromberg & Cohn, 2008) has shown that both M85 and E85 in a low-displacement, spark-ignited, high-compression gasoline engine can obtain efficiencies exceeding that of diesel engines. So, in effect not only does this eliminate the near-factor-of-2 calorific (mileage) penalty of methanol, but you end up with a smaller engine that gets better mileage than a gasoline equivalent. Also a higher-torque, sportier car to boot!

Heavy-Duty Vehicles

This class of vehicle uses of 2.5 MM bpd of oil, out of a total 8.1 million barrels imported each day. Consequently it is a logical target. FedEx recently announced an intention to switch its trucks to LNG. MIT research has shown (Cohn, 2012) that a direct-injected spark-ignition engine can be made smaller and more efficient. A typical 15-liter-displacement engine can be replaced with a 9-liter engine. This utilizes the high octane rating of methanol as well as the evaporative cooling mentioned earlier. The engine weight can be reduced from 3,400 pounds to 1,800 pounds, and the exhaust control can be simplified. However, the fuel tank needs to be 300 gallons instead of 200 gallons. But the added weight of this feature is completely offset by the lighter engine.

The High-Efficiency FFV: How Do We Get There?

Any large industry such as the automotive industry is slow to change, and with good reason. The original ethanol-based FFV was assisted with a federal subsidy to cover the engine modifications. If the broad mandate for FFV capability on most new cars goes through and if methanol production gears up, the compelling economics will cause consumer demand. This will create fueling infrastructure, perhaps assisted by one-time subsidies. At this point the risk of introduction of a high-efficiency FFV is minimal. By all accounts the automobile cost will be lower than for the gasoline counterpart of equivalent performance. This is in part due to the smaller engine and the more inexpensive emissions control equipment.

The automobiles will appeal to the buying public for their defining characteristics of a small, high-performance engine with long range fueled by a cheaper alternative to gasoline. The feature of nearly half the carbon dioxide emissions per mile as compared to an equivalent gasoline engine will also appeal to some. This last comes about from the fact that methanol has about half the carbon of gasoline, which causes the mileage penalty in an ordinary engine. But the high compression feature simply makes the engine more efficient, in effect negating the low carbon penalty.

An interesting entrée could well be through the military. On the one hand, they should have an interest in more efficient fuel utilization to minimize fuel transported to the front lines. On the other, methanol production at major bases using the feedstock of convenience would not be out of the question. A civilian version of a military vehicle could then follow, much as happened with the Humvee (originally the High Mobility Multipurpose Wheeled Vehicle, or

HMMWV). Ironically, the high-efficiency FFV (Heffvee) would be almost the opposite of the gas-guzzling Humvee.

When Nobel Laureate George Olah and colleagues proposed the Methanol Economy (Olah, Goeppert, & Surya Prakash, 2009) first in 2006, they were a bit ahead of their time. Shale gas had not made its presence felt to assure a long-term future of moderately priced gas. Now the promise can be realized. While methanol will be the main driver in gasoline replacement, uniform acceptance of first FFVs and then Heffvees will allow for ethanol to be used as well. Gasoline will take its place as just another fuel, not the dominant one.

The Road to Energy Independence

Responsible production of shale gas will essentially eliminate import of natural gas. That leaves the big ticket item—oil. Here, too, the notion of independence can usefully be defined as independence from distant and unreliable sources. The first step could be to target the oil passing through the Strait of Hormuz. Iranian saber rattling today concerns that flow.

The EIA forecasts that in 2022 we will import 7 MM bpd, down from the 8.1 MM bpd in 2011 (EIA, January 23, 2012). I think that if pipelines are built from North Dakota, oil from the Bakken formation will eat into this number more than already forecast. But sticking with their figure, first subtract imports from Canada and Mexico. Canada can be expected to ramp up its current flow of 2.2 MM bpd to at least 3.0 MM bpd. We have a special relationship with the Canadians: the bulk of their oil can only be refined in the US. Aside from the high carbon footprint of this oil, this is a desirable and secure relationship. Mexico currently supplies 0.8 MM bpd. This is at considerable risk of decrease but we will leave it at that figure for 2022. This, too, is heavy oil suited to our refineries.

Of the 3.2 MM bpd balance, I estimate about 1.7 MM bpd passing through the Strait of Hormuz. Therefore, one strategy would be to target oil alternatives to that level. Ignoring for the present the fact that a barrel of oil does not generate a full barrel of transport fuel, we can target 1.7 MM bpd of oil replacement. A rough calculation of all sources indicates this is viable, as enumerated below:

- Sasol has already announced construction of a GTL plant in Louisiana reportedly rated at 96,000 bpd of fuel. Assume at least one other such, bringing the total close to 200,000 bpd from GTL emboldened by low gas prices.

- In my chapter on Alaska (chapter 13), I suggested means by which at least 200,000 bpd capacity could be added to the pipeline. Absent some such action, the 500,000 bpd currently sent down from Alaska is at risk due to pipeline economics.

- Long-haul trucks switching to LNG or methanol could reasonably target 20 percent of current fuel usage, which accounts for 0.5 MM bpd of oil.

- Methanol, ethanol, CNG, biofuel, and electric cars could target 1.0 MM bpd. A significant part of this, and relatively straightforward, would be CNG displacement of diesel in metropolitan public and commercial transport.

An angle other than a Strait strategy is a study of the marginal domestic barrel and what it replaces. New domestic oil production is all light sweet oil. This is most like the oil from Saudi Arabia and Nigeria. So that may make sense as the first to be displaced. The Saudi portion is, of course, Strait-related and currently stands at about 1 MM bpd. Similarly, the uptick in Canadian oil that I predict will displace heavy crude such as that from Venezuela, currently about 0.65 MM bpd. The main point is that crude quality is variable and refineries are choosy, so country strategies have to recognize this.

Shale gas produced responsibly will be a key enabler for methanol to be produced at prices attractive with respect to gasoline. Broad availability of FFVs and associated fueling infrastructure will give the public choice. Tomorrow that choice could include other alcohols or methane, and a suggested high-performance FFV will enable that. Today methanol appears to be particularly advantaged. Ultimately, gasoline (and diesel) can be rendered just another player, not a champion. Game, set, and match.

PART IV

Informing
on Policy

CHAPTER 17

Turning the Pennsylvania Two-(Mis)Step into a Waltz

"There should be sunshine after rain"
—From "Why Worry" by Dire Straits (Written by Mark Knopfler)

There is an old adage: the people in the front get shot. Pennsylvania took some bold steps in the development of shale gas. They allowed shale gas production and the associated fracturing process. New York put a moratorium on fracturing and decided to study the matter. France and Switzerland did much the same. Industry-sponsored studies estimate that in 2010 Pennsylvania added $11.2 billion in economic activity, generated $1.1 billion in state and local taxes, and supported nearly 140,000 jobs. Industry opponents argue that groundwater was polluted. While each of these figures and assertions can be contested, and they are, they have underlying truths.

Fracturing operations have been routine for 60 years. Fracturing-enabled production of gas, and to a lesser extent oil, has been commonplace in Texas, Louisiana, Arkansas, Oklahoma, Colorado, Wyoming, Montana, and the Dakotas. So why the fuss about Pennsylvania, the birthplace of oil in this country?

The origins of oil in the US notwithstanding, Pennsylvania is largely unused to this sort of activity. The prospects lying in populated areas were also a factor. But the biggest impact likely came from the fact that the safe and cheap disposal of fracturing water was not practical here. In those other states, EPA-certified UIC Class II deep disposal wells were commonly available. The geology of Pennsylvania militated against this. In some cases the measures taken were woefully inadequate, such as sending the contaminated water to municipal disposal sites, which were ill-equipped to handle this waste. Others allegedly dumped the water. Yet others paid a lot to haul the water to deep disposal sites in Ohio.

The state is now taking stock of the substantial pluses and minuses. The actions will be watched closely by late-to-enter states such as West Virginia and Ohio. The ones on the sidelines (such as New York, Maryland, and North Carolina) will also watch with interest. Sometimes it pays to be a fast follower.

Making a Virtue of Being Late

This statement has the makings of an oxymoron. In many settings it certainly is. For example there can be no discernable virtue in being late for your own nuptials. Being late for one's own funeral, if that could be pulled off, has decided good points.

Being late is not the same as coming in second. Nobody knows that Tom Bourdillon and Charles Evans were within 300 feet of the summit of Mount Everest three days before the second team of Edmund Hillary and Tenzing Norgay got to the top. Bourdillon and Evans likely did not even make it into Trivial Pursuit.

In the business of innovation there is a body of literature on the value of being first. "First mover advantage" is firmly in the business lexicon. But so is the "fast follower" principle. Indubitably, fast followers could be faced with patents preventing their success. Intel went out in front early and never was materially threatened by Advanced Micro Devices (AMD). But many businesses have been built on the premise of letting somebody else develop the market and make the mistakes.

What does all of this have to do with energy? The history of development of shale gas is instructive. After the realization that horizontal wells and fracturing enabled gas production from these tight rocks, the early attempts employed methods previously used. In particular these involved using sugars as thickening agents to more easily fracture the rock. The sugar residue impaired production. Newer techniques, in areas such as the Marcellus, use "slickwater." The results have been dramatic, albeit at the expense of higher volumes of fresh water.

All of the foregoing is just plain building on the experience of the past. This essay on the virtue of being late keys on the point that if fate has dealt you a hand that causes you to be late to the party, find ways to make that a positive. This is the opportunity presented to the areas of the East Coast that have not yet materially been swept up by the shale gale. These include Ohio, West Virginia, Maryland, and North Carolina. These states must institute measures whereby the exploitation of the resource is done in an environmentally sound fashion while still maximizing the realization of economic value for the communities affected. The Grand Experiment in Pennsylvania will be highly instructive.

Taxing Natural Gas Production

We all recollect the now-famous phrase spoken by then-presidential candidate George H. W. Bush: "Read my lips: no new taxes." The public simply abhors new taxes and the politicians who propose them. At the time of this writing, House Bill 1950, the Marcellus Shale legislation, had just been passed by the Pennsylvania General Assembly. The major provision is an "impact fee" to be levied by each county. It would be proportional to the market price of the gas and have a time limit of 15 years. Understandably, the governor is loath to call it a tax.

While this appears to smack of politics-laden semantics, the suggested construct makes impact fee the accurate term. The levy is made by the counties and the proceeds are shared by the local municipality, the county, and the state. This allows for the money to be used explicitly to offset negative impacts of the industry, such as road damage. If targeted exclusively to this, not only would the term be completely accurate, but also the levy fair with respect to the industry. The state share may well be used in part to further job-related programs.

The public is more likely to accept gas production if it realizes that some of the proceeds are being used to benefit the community. This is especially the case for the majority who do not profit from lease royalties and yet suffer the traffic and other production-related inconvenience. In some ways this impact fee would stand in stark contrast to how the externalities with respect to coal and imported oil are largely given short shrift.

Pennsylvania is the only major gas producing state without a severance tax on hydrocarbons. This is basically a tax on either the volume of gas or its value and often has limited or total forgiveness for certain situations. As in the case of Pennsylvania's suggested impact fee, these tend to have sliding scales based on gas value.

Environmental Protection Measures

The report of the Pennsylvania governor's Marcellus Shale Advisory Commission is also instructive to any other states wishing to learn from the Pennsylvania experience. It covers a lot of ground, some of which is reflected in the discussion in chapter 24, "Policy Directions." At this point it is merely a guide to action by the governor and state government. But regardless of when the recommendations turn into law, they are available for other states to peruse and act as needed.

An important provision is the requirement to have baseline testing of proximal water wells prior to drilling activity. In addition to requiring this, the report stipulates that the operating company perform the task. They are also required to use an independent testing laboratory that is pre-certified by the state. The Pennsylvania Department of Environmental Protection (DEP) will post a list of state-certified testing labs to facilitate the process. A variant of this process could be a requirement for notification of proximal homeowners when a drilling permit is granted (this is in the DEP's recommendation now) and county assistance for proximal homeowners to get the baseline well testing done at the expense of the operating company. Among the advantages of this method would be the direct involvement of the homeowner in both selecting the testing entity and keeping track of the results.

An interesting recommendation is to double the civil monetary penalties for violations of Pennsylvania's Oil and Gas Act. The DEP is to be given authority to act in this matter and will generally be given a bigger stick for noncompliance, including suspension or revocation of permits. In general, so many of the needed regulations on production-related care and diligence are just good business for the operator. Punitive action ought to rarely come into play. This should reduce the cost of enforcement. Also reducing the cost would be the effective use of the latest technology in monitoring and transmission of the data to central locations at the oil and gas company. Remote monitoring and control of well operations are coming into vogue and would facilitate this regulatory purpose. This would largely negate the need for DEP physical presence on the rigs, avoiding a good deal of cost in the process.

Workforce Development

The state of Pennsylvania recognizes the need for thousands of trained workers in coming years. This is proving to be something of a challenge in some communities. The companies providing drilling and completion services provide their own specialized training. But base levels of qualifications are required depending upon the job. For many jobs, the required training can be as little as an associate's degree. Others require a bachelor's degree, preferably in technical disciplines. But almost all jobs face some common features: the hours are long (often unpredictable) and the work is hard. Furthermore, workers trained by these companies may be required to operate in other states and sometimes other countries. While the employers will be motivated to keep the personnel close to their original homes, the nature of the oil and gas business is one of a commuting life in many instances.

The Cranberry Effect

Cranberry Township in Pennsylvania has seen immense revenue growth due to the shale gas boom. No shale gas wells have been permitted or drilled in the entire township. What they did was create an atmosphere that attracted shale gas players, which located regional offices there. Support personnel such as accountants, lawyers, repair and maintenance outfits, and other professionals followed. It was simply a great place to live, work, and be entertained.

In some ways earlier history with another industry gave Cranberry an advantage. The town had succeeded in attracting a major research and development complex of Westinghouse, with nearly 5,000 employees. To do this, they developed the type of support services attractive to moderate- to high-income families. When shale gas drilling commenced all around them, they successfully targeted the operating entities. Today, the regional branches of several of the players in the Marcellus are housed in Cranberry.

This ability to get the gain without the pain I dub the "Cranberry Effect." They could in fact take the Effect to the next level by encouraging the downstream processing to be done in the township. One target could be the synthesis of anhydrous ammonia from methane. As noted in chapter 11, when methane is cheap this is a high-margin business. The other would be relatively small-footprint ethylene crackers along the lines suggested in chapter 12, "The Ethane Dilemma."

The town's close proximity to the ethane from wet gas would make the location particularly attractive. These jobs are high paying and tend to have well-scheduled shifts, unlike the drilling business. The work would also not be considered hard in comparison with other industrial jobs (some of the reluctant drilling job applicants had given this as a reason for recalcitrance). Most importantly, workers would be hired for that location only and not be required to spend time in other states. In other words, none of the objections to the drilling jobs would be applicable here.

The unusual time patterns can be a severe handicap to family life. It is less so for young workers without families, and seniority usually brings with it more regular hours. Studies among uranium miners in West Virginia have identified behavioral problems caused by the nonstandard working hours. In some cases the incidence of alcoholism went up.

In Butler County, unemployment numbers in 2010 were at decadal highs. One would have expected a high level of interest in these jobs. This has proven to largely not be the case due to the expected working conditions and has caused the authorities to take two approaches. One is to focus on attracting midstream and downstream companies to the area. Midstream includes aspects such as pipelines and associated pumping equipment. Downstream is defined largely by activities such as gas processing and conversion to higher value products. Shell has already announced an ethylene cracker for the region. The second approach to jobs is to attract regional headquarters of the operating companies. The township of Cranberry has had success with this angle.

Other Societal Issues

True adherence to the principles of sustainability would require that the community not only not be harmed but that it actually accrue some of the benefits from industrial enterprise. A case in point is the entire state of Alaska. Every resident receives a dividend check from the state representing a portion of the government revenue from the oil business. Families in Texas in effect receive a dividend: there are no state income taxes.

These examples will not necessarily find equivalence in other states because much of the revenue likely is from leasing of state land. Also, these payments are cash transfers, which are welcome by residents, of course. But it is arguably time to better integrate the benefits of science, technology, and industry within the nearby communities that have natural gas deposits. An example could be creating research stations relevant to the specific technologies used for extraction and transportation. Is there a way to create more positive spillovers for locals so the benefits are felt over time? These are complicated prospects, but fresh thinking on ways to do this may make natural gas extraction and fracturing much more palatable to communities.

Pennsylvania took a bold step. All the evidence points to them waltzing into prosperity provided the important elements of the governor's Commission recommendations are implemented. Others will learn from this, but Ben Franklin's state is likely to capture the first-mover advantage.

CHAPTER 18

Will Cheap Natural Gas Hurt Renewables?

"The answer, my friend, is blowing in the wind"
—From "Blowing in the Wind" by Bob Dylan

Yes, cheap natural gas will hurt the rate of growth of renewable energy. There is no way to sugarcoat this unfortunate outcome. But much can be done to ameliorate the effect, and most of that lies in the policy arena.

Ironically, the celebrated successes of NGOs, primarily the Sierra Club, in shutting down coal plants and halting the building of new ones may hurt the cause of renewable energy. In the time frame required, in many instances wind and solar options will not meet base load grid parity. This is generally defined as parity with the cost of base load electricity production in that area. So, absent strict policy measures, which will be very hard to come by, natural gas will be the fuel of choice. Natural gas plants will not simply be mothballed when wind power reaches economic parity. The early demise of coal plants will lead to natural gas-fueled electric power plants with greater installed capacity.

In recognition of this, the Sierra Club had initially taken a position of natural gas as a bridge fuel until renewable sources became more economically viable. Hand-in-hand was the insistence on environmentally responsible production. In early 2012, Sierra Club executive director Michael Brune seemingly reversed that position, citing the environmental risks posed by shale gas production. He said, "As we phase out coal, we need to leapfrog over gas whenever possible in favor of truly clean energy" (Brune, 2012). While the "whenever possible" phrasing leaves some wiggle room, this is a definite movement away from the Club's original position.

In the view of this author, a life member of the Sierra Club, the solution is the responsible production of shale gas as bridge fuel. Clean coal is not an oxymoron, but it is expensive. At the same time, the move to renewable energy must be accelerated. In another odd twist, natural gas may be needed at first as a load leveler, even for wind and solar production, due to their diurnal cycles. Someday we hope to develop effective storage mechanisms. But someday can be a while, and we must be realistic regarding the need for bridging mechanisms. The 2012 Department of Energy budget includes funding for an Energy Innovation Hub researching energy storage. This is good policy.

The price of natural gas in North America is roughly one-fourth that of $100-a-barrel oil on the basis of energy content. For this computation, I consider a more normal $4 per MMBTU, not the $2.50 in March 2012. In Europe the price is higher but still a factor of 2 to 3 cheaper than oil. The discussion of whether renewables will be hurt falls in two distinct realms: electricity generation and transport fuel production. Accordingly, I consider these separately because the impacts will be distinctively different.

But first, let us examine the very premise of cheap natural gas. Until the shale gale swept us up, natural gas prices fluctuated considerably. For the last 13 years, the Henry Hub price was as low as $2 per MMBTU and as high as $13 (Figure 12). The Henry Hub price is the price of gas in Erath, Louisiana, and is used as the price for trading on the New York Mercantile Exchange. It is a rough proxy for US natural gas price.

Figure 12. The Henry Hub price per million BTUs, 1999–2012

Source: US Energy Information Administration, May 9, 2012

The inherent uncertainty in gas price caused coal and nuclear to remain as viable options for electricity generation. Entire chemical industries moved abroad to regions of predictably cheap gas.

Gas from shale at first was costly. After the kinks got worked out and further advances were made, it could be produced profitably at costs lower than many conventional gas operations. Meanwhile, the sheer volume kept the price down. At the prices today the most profitable operations are those with higher proportions of natural gas liquids in association, because the value of this component is pegged to the price of oil and benefits from the high price of oil. The dry portion of the gas therefore continues to be produced, even at low prices.

Inevitably, consistently low prices will create demand and eventually prices will rise. In the face of this I offer the view that shale gas production will keep gas prices moderate. This is largely due to shale gas wells being on land and shallow by industry standards. These wells can be in production in 30 to 60 days after commencement. This short duration effectively keeps a lid on the price. If the three-month strip (the commodity price three months into the future) is seen as going up, new wells can be in production well within three months. This sort of certitude will also discourage speculative investment in the commodity.

Effect on Electricity Production

The floor price will get set by the conversion from coal to gas for electricity. Forty percent of coal plants not expected to meet the latest EPA standards on mercury and NOx are over 50 years old; these fully depreciated plants will not be refurbished. The only options are new coal, nuclear, and natural gas plants. New coal is disadvantaged on price alone until natural gas reaches a price of $8 per MMBTU. Recently that price has been under $4. So, with the ceiling mentioned in chapter 3, coal is not the economic choice. Nuclear has suffered a blow due to the Fukushima Daiichi disaster. So, natural gas will be the fuel of choice. Eventually, the shift to gas could cause the price to rise, but the lid will still be around $8. A wild card on price is the possibility of strong liquefied natural gas (LNG) export demand. This is not far-fetched, especially to Japan, where natural gas prices are about triple those in North America. The total added cost of liquefying, transporting, and re-gassing natural gas would be about $3 per MMBTU. This puts the landed cost at well below the market price in Japan. However, such export would require federal permits which may not come easily.

Cheap natural gas will also cause a shift from oil to gas whenever possible. This additional demand will keep the price up in the medium term. So, let us assume a price of $8 as the ceiling price. At this price, electricity will be delivered at a little under 7 cents/kWh. This is the price that alternatives will have to meet on a direct economic basis.

This natural gas–derived electricity price is lower than the fully loaded price of energy from new nuclear plants, which will be over 10 cents. Currently wind delivers at 9 to 16 cents depending on where it is. Offshore wind may be higher yet at this time. Wind also often suffers from the need to add transmission infrastructure. This is especially the case for offshore facilities. There is also the celebrated case of T. Boone Pickens' terminating a major land-based investment due to the absence of definite plans to add transmission lines.

Strictly from a techno-economic standpoint, wind still has an upside. Engineered solutions are likely to drop the price from current levels. But it continues to suffer from diurnality, and so it needs to be companion to another source or to storage mechanisms.

Effect on Transport Fuel

Transport fuel today is dominantly produced from oil. Oil prices can be expected to remain high, in part due to the burgeoning demand from India and China. This demand is a direct result of the sustained increase in per capita GDP in India and China; a strong correlation exists between this metric and per capita vehicle ownership. On the face of it, this prediction strongly favors gasoline and diesel substitutes derived from renewable sources.

While only a few today, such as Brazilian ethanol, have achieved parity on a cost basis, there is room for improvement. This is particularly the case for drop-in fuels from sugarcane or sugar beets. Butanol is an example of a liquid that is a complete gasoline substitute in any proportion, unlike ethanol, which is more corrosive and water absorbent, and suffers a calorific penalty—that is, it has a third fewer calories than gasoline and commensurately worse fuel economy.

However, one does need about 8 percent ethanol in gasoline simply to act as an oxygenate, which makes the combustion of gasoline more complete. The chemical MTBE was previously used for this purpose, but has since been outlawed for environmental reasons involving pollution of groundwater.

Currently the most promising avenue for drop-in fuels is production of alkanes from sugar. The candidate raw materials are sugar cane, sugar beet, and sweet sorghum. Alkanes are straight-chain compounds with the formula

C_nH_{2n+2}. Conventional oil-derived fuels have this formula as well. The number n is about 7 to 9 for gasoline and about 12 for diesel and a bit higher for jet fuel. So, alkanes with the right number are for all practical purposes direct drop-ins for these conventional fuels.

Herein lies the attraction. Also, being tailored, often through genetic engineering, the composition will be predictably uniform. This is not the case for the input to refineries from a variety of crude oil sources. In fact oil refineries today are forced to be very picky about the mix of crude they will accept. Seed-based oils also suffer from this variability.

No small wonder, therefore, that many of the leading players in the drop-in biofuels space are supported by major oil companies such as ExxonMobil, Shell, and Total—all heavy hitters.

Cheap natural gas can affect transport fuel from renewables in two ways. One is the direct use of methane as fuel for combustion in the engine (see chapter 12); the bottom line is that methane is unlikely to be a material factor for passenger vehicles. The other is the production of transport fuel from conversion of natural gas, known as gas to liquids (GTL; see chapter 14); the point of discussion here is that cheap natural gas may be converted to transport fuel for a relatively low cost. If that cost is low enough it could disadvantage the schemes described above.

However, the proven technology in this arena, Fischer-Tropsch synthesis, is still expensive. Further research in catalysis will reduce its cost. Right now no compelling evidence suggests that alkanes from sugar will be particularly disadvantaged with respect to GTL, even from cheap methane. Finally, methanol from methane is a wild card, as discussed in chapter 16. Another wild card is methanol from biomass.

In the end, the true yardstick will be comparison with the price of imported oil, not with other oil alternatives. Gas-based oil substitutes could coexist with renewable liquid fuels.

Policy Matters

Without a price on carbon, the carbon-free alternatives wind, nuclear, and solar are seriously disadvantaged. Taxes are anathema to Congress. Cap and trade has not worked particularly well in Europe, in part due to the uncertainty, which effectively increased the discount rate on investment. Also, any cap and trade conceived by Congress will undoubtedly have numerous exclusions and grandfathering. The province of Alberta in Canada has an interesting model. They tax high-carbon-footprint heavy oil production over a certain volume. The rate is $15 per tonne ("Go Figure," 2007). The money is placed in a special fund expressly for the purpose of addressing environmental issues associated with oil and gas. Such directed use of tax proceeds is more palatable. Conceivably, the fund could subsidize renewables for a period of time.

Finally, one could resort to the current method of imposing a renewable portfolio standard, which essentially mandates that a proportion of delivered power be from renewable sources. This in effect is a tax on the consuming public because the renewable energy costs more. The solar subsidy in Germany is passed on directly to the consumer as well. But that is largely possible due to the considerable influence of the Green Party. Short of taxing conventional oil and gas, consideration could be given to decreasing the incentives and redirecting those funds. But policy, if not based on an understanding of market forces, can have unintended consequences. An example of that is the "Flex-Fuel Fairy Tale" (chapter 22).

Conclusion

Cheap natural gas will place every other source of electricity production, including renewable energy, at a disadvantage for the short to medium term. Reliance on market forces alone will slow the introduction of renewable energy. Policy mechanisms are needed to level the playing field, at least from the standpoint of carbon neutrality. The most equitable methods may be a US analogue to the method used in Alberta. By all accounts that policy is embraced by the public and industry alike. If the cheap-gas-enabled methanol displacement of gasoline and diesel becomes a major factor, even biofuels could be negatively impacted. On the plus side, methanol could be synthesized from biomass.

CHAPTER 19

Kicking Shale into the Eyes of the Russian Bear

"You don't tug on Superman's cape"
—From "You Don't Mess Around with Jim" by Jim Croce

On January 7, 2009, Russia shut off the natural gas ("Europeans Shiver," 2009) flowing through the main European pipeline in the Ukraine. This was a particularly cold winter and 20 European countries encountered serious natural gas shortfalls. Discussed below are the reasons given by all of the players. But the principal point was, and continues to be, that Russia can use natural gas supplies as a weapon to achieve political objectives. In late 2008 Russia threatened to form a gas-based OPEC (dubbed OGEC) with Iran and Qatar with the express intent of manipulating world gas prices. Has shale gas dampened their ardor?

Unilateral fuel cutoff as an instrument of political will would be essentially not possible with oil. Oil is more fungible, and alternative supplies can be brought to bear if a major supplier falters, deliberately or otherwise. It may cost more but you could get it.

Natural gas is a regional commodity. Bulk transport across land can only be done through pipelines, and these are expensive and have long lead times. Transport across the ocean is feasible only if the gas is liquefied. (For shorter distances there are exceptions, where gas pipelines cross bodies of water, such as the North Sea.) The product is known as liquefied natural gas (LNG). This process entails cooling the gas to -160°C into a liquid that is 600 times as dense as free gas. This is then transported at close to atmospheric pressure. The low temperatures are maintained by auto-refrigeration through allowing small amounts to boil off, which causes chilling of the remaining liquid. An everyday analog is cooling of our skin by a fan or a breeze causing evaporation of our perspiration.

While LNG is a viable alternative to a domestic gas supply, it can only be delivered to a port location, and in fact only one with a re-gas terminal. The high capital cost of such terminals is unlikely to justify a capability merely to be available for upset conditions. So, as a practical matter, withholding of a domestic source is a powerful weapon, LNG alternatives notwithstanding. Also, LNG is more costly. Typically the added cost over the price of the gaseous version is about $3 to $4 per MMBTU. (Transport distance is the determinant of where you are in that price range.) As a frame of reference, that is roughly the price of natural gas in the US today. So LNG would essentially double that. This is why cheap shale gas in North America has rendered imported LNG passé.

The sheer distance between producer and user is the reason natural gas prices are so variable across the world. The price in Europe is about double that in the US, and in Japan the price is about triple and even more, as seen in the Figure 13. The latest figures are likely impacted by the Fukushima Daiichi disaster-driven shift to natural gas for power. Depending upon the Japanese government's action with respect to future nuclear plants, the price could remain high. In general the high prices are in part because costly LNG is the marginal cubic foot, and so sets the price. "Marginal cubic foot" is industry terminology to mean the last tranche of gas added to serve demand. Since it is essential, the high price gets paid. This inevitably increases the price from all sources because the domestic producers can charge that figure. The

Figure 13. Trends in natural gas spot prices at major global markets

Source: Energy Information Administration, September 30, 2011

LNG business is built on long-term contracts because the capital cost is in the neighborhood of $4 billion. Nevertheless, there is considerable arbitrage; tankers will sometimes reroute in open waters to new destinations offering higher prices.

Russian Use of Gas as a Weapon

Unlike in the Soviet era, Russia can no longer impose its political will through threatened military action. However, Russian gas is a significant natural gas source for most European countries. It is the dominant source for nine countries, including Greece, Finland, Hungary, and the Czech Republic. This monopoly allows unilateral action against any one of the countries. Action against too many would result in loss of needed revenue. (The Arab oil embargo in 1973 had a profound and lasting effect on the price of oil, aside from the short-term privation. But the original political objective was not realized, that of causing a significant shift in support away from Israel. Interestingly, though, the lasting price escalation that was a direct result of the embargo swelled producing country coffers. This allowed financing of politically motivated actions in other countries, including the funding of Islamic schools known as madrasas in Indonesia and other countries. These are believed by some to be linked to militancy.)

In an odd twist, the embargo-driven sustained higher prices opened up exploration in promising but costly areas such as ultra deep water and the Arctic, thus reducing dependency on OPEC. Since then Norway and Brazil have become important players, on the backs of deep water development.

The Russian action in 2009 was allegedly driven by a dispute with the Ukraine with respect to poaching on the gas line. While there may have been truth to this, most believe the action was intended to injure the Ukrainian Orange Revolution, which was seen by Russian President Dmitry Medvedev as not commensurate with Russian interests. The temporal connection strongly implies causality with the gas cutoff action. In many ways this act was more effective than would have been a military one. It also undoubtedly sent a message to other European states. Even Western Europe was affected, with southern Germany losing about 60 percent of its imported gas.

Shale Gas Could Change That

As discussed in a chapter 1, the mechanism by which shale gas accumulates makes it likely to be ubiquitous. So the likelihood of substantial deposits in Europe is high. Initial estimates by the EIA show large deposits in Poland and France, with smaller amounts elsewhere, including the UK and the Ukraine. Poland is actively exploring and the UK is following suit. France currently has a moratorium on fracturing, but is also not as much in strategic need due to low dependency on coal-based power. US efforts to produce gas with a minimal environmental impact will be important for widespread exploitation in Europe. Poland is certainly resolute on the matter. Furthermore, as in the US, as exploration proceeds, the resource estimates are bound to increase. All new hydrocarbon resource plays follow that pattern.

Gazprom, the mammoth Russian company operating gas assets, has publicly expressed concern regarding the effect of shale gas on future pricing. The fact that Russia too will have large deposits is irrelevant. Further increase in their resource base is interesting, but not a factor in the concern regarding domestic sources in client countries.

An interesting possibility is that US shale gas could be exported as LNG. Until European deposits are developed, US-sourced LNG could be a factor in offsetting Russian supply. If US prices remain low, as is expected, the cost of delivered LNG in Europe could profitably be at under $9 per MMBTU for some years and closer to $7 today. Furthermore, major LNG developments in Qatar and elsewhere which had been destined to supply the US will now find Europe a ready buyer. From a Russian standpoint this will not be a pricing concern, but certainly the gas as weapon argument is affected. Strictly from an economic perspective, the best sources for North American LNG are gas from Alaska and British Columbia, and the most logical target customer is Japan because of location and price.

OGEC Is Dead

Sixty percent of the conventional gas reserves reside in Russia, Iran, and Qatar. Operating costs are very low, especially in Iran and Qatar. In late 2008 the three announced an intent to form a gas-based OPEC, which was dubbed OGEC. (Note: The *P* in OPEC is for Petroleum, and by definition, albeit not by common usage, gas is included in the term *petroleum*, so the acronym OPEC could have applied to gas as well in theory, but for the existing different cast of characters that would not have made sense.) Alexey Miller, chairman of Russia's Gazprom, said they were forming a "big gas troika." He also predicted an end to the era of cheap hydrocarbons, thus signaling the intent of the gas cartel to raise prices and keep them high. OPEC accomplishes this despite supplying only about a third of the world's oil.

The troika would likely have been pretty effective, in part because Russian markets are Europe and China over land, and Iran and Qatar are much more LNG-dependent. So, unlike current OPEC members who compete in the same markets, at least the senior partner, Russia, would be essentially noncompetitive with the other two, except for LNG relief valves for Russian force majeure, contrived or otherwise.

Shale gas over time will kill attempts at OGEC. China is expected to have even more shale gas resource than the US and will exploit it quickly. China National Offshore Oil Corporation (CNOOC) has already taken ownership positions in two US shale gas development projects and in the first large one in the UK. (There is little doubt that part of the intent is to transfer technology to China deposits.)

European shale gas will certainly be a factor. There is reason to believe most of the countries currently importing LNG, including India, have shale gas opportunities. Finally, there is the specter of US as an LNG export player. All of this adds up to a world with a lot of gas in consuming countries and more options. When consumers have options, cartels are ineffective. Gas has always been harder to manipulate than oil. Transportation needs can only be met by oil-derived products. Gas on the other hand can be replaced by coal, wind, and solar for power. OGEC can be pronounced dead on arrival, and we have shale gas to thank for that.

Shale Gas and US National Security

"Oh peace train take this country"

—From "Peace Train" by Cat Stevens

The biggest winner from a prohibition of fracturing and hence shale gas production would be Russia, closely followed by Iran. The ability of Russia to impose its political will upon the world, and Europe in particular, will be significantly enhanced. Absent shale gas, Russian gas will increasingly enjoy a monopoly in many European countries. Gas supply throttling could be used as a weapon of political will, as it already was once in 2009. Although the target was the Ukraine, the Slovak Republic suffered enormous collateral damage, estimated by some to be €100 million per day for the 10-day period. This caused gas-deprivation-related recession. While much of this effect is attributed to unpreparedness, gas supply as a weapon is clearly real.

In such a scenario, Russia would also have an increased ability to side with countries with interests opposed to those of the US. In particular this could apply to not joining in sanctions or other measures against Iran. Consequently, Iran would be an indirect beneficiary of this power.

Iran would benefit more directly from a world without shale gas. Prices for gas will rise, and almost as importantly, will be unstable. Internal squabbles have delayed bilateral deals with countries such as India and Pakistan. Sustained shale gas production in the US will close this country as a destination for LNG, thus closing a window for Iranian LNG exports in general. The LNG from Qatar and elsewhere originally destined for the US will find other markets that Iran may have targeted. In fact today, with shale gas accounting for about 30 percent of US domestic production, LNG contracted to be delivered to the US is being delivered elsewhere, including India.

India is perennially short of hydrocarbons. Absent shale gas, and possibly even with it, India will rely heavily upon gas from countries such as Myanmar, Iran, and Qatar. The first two have the potential to be pipeline services, although each is beset with the issue of traversing nations not friendly to

India. That leaves LNG from Iran and Qatar. Any pipeline would be very expensive, especially if it follows an ocean route, as would be the case for an Iranian pipeline avoiding the Pakistani land mass. This would mean long-term contracts. Shale gas–induced lower pricing would dramatically reduce the pricing on these contracts, with attendant beneficial impacts on the Indian economy. Today India is importing gas at about three times the US price.

Any Indian dependency on Iran will strain US relations even without the current (2012) embargo conditions related to nuclear weapons. One could extrapolate reduced US influence in the subcontinent as a result. The US-Pakistan relationship is already in some strife, and the US can ill afford this additional complication in the subcontinent.

Reemergence of OGEC

The Organization of Gas Exporting Countries (OGEC), comprising Russia, Iran, and Qatar, will certainly have the ability to be resurrected if shale gas production is curtailed. Around 2009, Russia was very open about using the cartel as a means to higher gas prices. Higher export prices for gas would put even more revenue into their pockets and those of Iran. The world has already seen the mischief wrought by Middle Eastern oil money following the 1973 Arab oil embargo–driven sustained higher oil prices. It would however be specious to suggest that OGEC-driven gas price rises would have similar effects, because gas simply is not as fungible as oil. But gas is a much more basic commodity in that it affects many walks of life, whereas oil is predominantly used for transport.

If LNG becomes a significant fraction of gas consumption, keeping shipping lanes open will now assume importance for this commodity as well. Military costs will for the first time constitute an externality for gas. Much as is now the case with oil, conflict in the Middle East could start having an impact on US natural gas prices.

Energy Security

The International Energy Agency defines energy security as "uninterruptable physical availability of energy at a price which is affordable, while respecting environment concerns" (International Energy Agency, 2012). From the standpoint of national security, energy security is an important part. While a physical threat to the populace in the form of attack is a part of national security, a large aspect is protection of the economic way of life.

Examining first the issue of affordability, domestic sources do not equate to low cost, at least in the case of oil, because it is a world commodity. Some countries create low prices through subsidies, but these often have consequences. In India, where kerosene is a way of life for cooking and other purposes, affordability is created with a subsidy. However, this is at the distributor level and so is subject to diversion of subsidized fuel to profit-making enterprises. Even were this not the case, this setup creates a burden on the taxpayer. Subsidies or inducements to use less ought to be at the consumer level. On the latter point, some commercial offerings are subject to a tiered pricing system, but this is not the norm for household power. With electricity it ought to be simple to provide the first tranche of power at low prices, thus ensuring essential heating and cooling for the less fortunate.

Affordability applies to industry in a big way. As discussed earlier, entire industries left the US when gas prices were high and variable. Oil does not elicit that reaction as much because there tends to be a world price because the commodity is fungible. Certain states have been able to attract industry almost solely by the low power costs in the region. A subplot in that case is that in some instances the average cost is low due to a preponderance of older, fully depreciated, coal plants. With the passage of time these will need replacement and costs will rise. The move to renewable energy will also raise costs. These effects can be dampened if shale gas–enabled prices stay low for natural gas.

The usual broad argument for drawing a bright line between energy availability and national security is that of support of oil-rich countries with dubious leadership. Protecting our way of life involves ensuring supply, which implicitly supports activity potentially adverse to the national interest. One of the positives associated with shale gas is that it affords us our way of life without swelling the coffers of the likes of Russia and Iran with gas revenue to create mischief.

The Baker Institute Study

A recent study by the Baker Institute (Medlock, Jaffe, & Hartley, 2011) in Houston describes a detailed analysis of the effects of shale gas on national security. They constructed three scenarios, of which I will analyze just two for simplicity: the base case of full exploitation of shale gas and the extreme case of no further exploitation beyond that in Texas and Louisiana. In each case they calculated the decadal averages for gas price for the next three decades. In the base case they showed the average price of gas at $5.84 and $6.46, respectively, for the 2020s and 2030s. These numbers are in general accord with my projections in chapter 3. But my key point is that not only will averages remain moderate, but the excursions will not be large. Stability is almost as important as low prices because this certainty drives investment. The reason the chemical industry is returning to the US is as much the certainty as the current low gas prices.

The Baker Institute study is probably a bit optimistic regarding gas prices in the scenario without widespread shale gas exploitation. The averages may be right, but the excursions, which they do not appear to model, will almost certainly be large due to reliance on imported gas and perturbations due to upset conditions in the exporting countries.

The Military and Energy Security Nexus

The military is the largest consumer of energy in the US public sector, consuming 5 billion gallons of fuel in 2010. Access is not really the issue even in times of tight supply. But it is incumbent on the military to reduce its reliance on fuel while at the same time not sacrificing operational effectiveness. This applies to all forms of energy, not just fuel for transport, although that is the one with the greatest imperative.

During the Iraq War, there was great deal of public angst regarding the price of fuel for the war effort, and many in the supply chain got blamed. The fact is that a captain in a forward emplacement is not worrying about the price when he or she needs fuel urgently. The monetary cost aside, the human cost of such delivery is substantial. In the Iraq and Afghanistan wars in 2007, an estimated 3,000 military and civilian support personnel were killed or wounded while transporting fuel or water. Reduction in fuel usage, substitution with more benign alternatives, local sourcing of energy and water—these all ought to be priority strategies for the military today.

The time was never more right than now to innovate in reducing the cost of energy in the military. The budgetary toll will be heavy this year once the congressional squabbling is over. One war has wound down and another is on the way to ending. The next war must be supported by low energy methodology, running the gamut of lighter vehicles (fits well with the smaller lighter army motif in vogue today), fuel replacement to minimize high-risk convoys of diesel and gasoline, off-grid distributed power with renewable energy supported by microgrids, desalination of saline groundwater to minimize water transport, and electric vehicles when feasible, because distributed power is lot easier than distributed fuel generation.

Semi-permanent bases domestically and abroad could even invest in distributed fuel production. If natural gas were readily available, small-footprint production of a drop-in fuel would not be out of the question. Given this possibility, the military ought to fund such developments rather than massive coal- and gas-to-liquids projects. In any case, small-scale distributed power in the form of mini-nuclear (what would be more secure than a military base?), wind, and solar combined with a micro-grid could power entire bases off the grid. This would not only give a green feel, but also would render the base relatively impervious to weather- or sabotage-related grid outages. Certainly in forward locations, the solar option would apply.

Base vehicles are uniquely suited to fuel switchover. This is because infrastructure support for refuelling is straightforward. Furthermore, in the example of CNG and LNG substitution of diesel and gasoline, where feasible the engines ought to be modified to take advantage of the high octane rating of methane, thereby delivering more power and distance for less fuel. The same goes for electric vehicles. Again, distributed electricity is easier and an electric vehicle delivers 60 percent more miles per unit of energy consumed. Not only will imported oil be substituted for, but less energy will be used. Only certain vehicles may be suited to electrification, but any gains would also have the virtue of symbolism.

Fresh water transport to front lines does not get much press but is a tractable objective for reduction. The shale gas industry will be learning to deal with low-cost water sourcing and treatment. These advances could be used to advantage by the military. Saltwater aquifers are fairly ubiquitous, and the shallower they are, the less salty. The Defence Department ought to consider sponsoring developments of small-footprint desalination plants, especially targeting the types of salt water anticipated in theatres of action.

Every president in the last decade or so, no matter from which side of the aisle, has drawn that bright line connecting energy security and national security. President Bush, a champion of oil and a onetime owner of oil interests, famously complained about our "addiction to oil." President Obama recently said, "America's dependence on oil is one of the most serious threats that our nation has faced." That sounds like a national security statement. So, equating national security to energy security and thence to reduction of imported oil will not be disputed by many.

Sustainable Development: A Double Bottom Line, Plus Afterthought

"For the times they are a-changin'"

—From "The Times They Are a-Changin'" by Bob Dylan

The definition of sustainable enterprises is the so-called triple bottom line, wherein economic, ecologic, and community benefit are all considered and balanced. Is that last leg of the stool given mere lip service, or is the energy production industry recognizing this element fully? And ought it to be?

The economic consideration is a given. Without that there is no profit, and absent profit, no enterprise. The ecologic or environmental piece is much in evidence today, and few new energy enterprises would dare ignore this element. The societal element is harder to define. One is tempted to think that this is strictly composed of negative impacts upon society, because that is where the rhetoric is directed. In some ways it suits detractors of the goal of sustainable enterprises to cast it in this light rather than a more generic one. So, for example, visual pollution is denigrated as a personal preference rather than as pollution in the classic sense.

The Reality of Visual Pollution

Perception is reality, the saying goes, and marketing folks well know that this is a powerful adage. One cannot bully people into feeling a certain way. Certainly not in commerce. But on an issue of alternative energy, some nudging, in the sense of Thaler and Sunstein, is in order. Richard Thaler and Cass Sunstein wrote a powerful essay, "Libertarian Paternalism," in the top economics journal *American Economic Review* (Thaler and Sunstein, 2003). Non-economists, such as I, must not be daunted by the staid prominence of the journal; this is an easy read. A further easier read, one that costs some money or trouble (going to the library) is their book *Nudge*. Basically they posit the notion that given free choice, people generally do not make the best decisions for themselves, even in an economic sense. People need to be given a nudge. My point is that just because folks feel a certain way about visual pollution does not mean they cannot be nudged to a different position.

One way to do that is to clarify the options. Until recently the Sierra Club was against coal, nuclear, and hydrocarbons in general. (Incidentally, coal is a hydrocarbon, but one challenged in hydrogen content, and most think of it as a different species, but it is not.) Last time I looked, that position was taken as tantamount to suggesting we switch off the lights. Wind and solar are great options. But they are still fledgling and incapable of base load service. In the interest of fairness, the Sierra Club now supports natural gas as a transitional fuel, still to the consternation of much of the membership. Even more lately that position has softened to one of "leapfrog over gas whenever possible in favor of truly clean energy" (Brune, 2012). This is a laudable sentiment that will be tough to execute in time for all the forthcoming coal plant closures brought about in part by Sierra Club activism.

Duke professors recently made famous by their paper (Osborn et al., 2011) connecting well water methane concentrations to shale gas production suggest in an op-ed piece in the Philadelphia Enquirer (Jackson & Vengosh, 2011) that we eschew shale gas in favor of wind and solar. No matter that each of these has opposition as well. There are entire communities that will not permit a visible display of solar panels on homes. Wind power has long been opposed on visual lines. North Carolina, the home state of the aforementioned professors, has a law preventing wind farms on mountain sites, known as the Ridge Law. Many communities have strong opposition to offshore wind production in sight of land.

When one flies into Amsterdam airport, one sees an abundance of wind farms in the water. Personally, I think they look like a flock of birds—but I am a techie, what do I know? Perhaps their acceptance is premised on the Dutch having had windmills as a way of life on farms. More likely is the explanation that it is a choice between that and Russian gas. In Holland that may not be the direct option, but in Greece, which is predominantly dependent on Russian gas, it would be. Southern Germany still has frigid memories of when the Russians capriciously shut down the pipeline through the Ukraine in the early days of January 2009 as a political move. So, opposition to something should come hand in hand with a consideration of the alternative.

Societal Benefit

Fair and equitable economic benefit to the local and regional communities ought to be a goal of sustainable energy development. In Australia's Northern Territories, uranium mining has provided a dividend to the native Aborigine community, conjuring up the image of traditionally garbed locals riding on the beds of Toyota trucks. Every resident of Alaska gets an oil-related dividend of substance. But these are the exceptions.

One measure would be similar to that in Alaska. Royalties on production would in part be distributed to the county in question. At the very least, this would go to ameliorate some of the damage to infrastructure. In the case of shale gas drilling, the principal damage coming to mind is the deterioration of lightly constructed farm roads by heavy trucks carrying sand or water. This will be a problem regardless of how well the water-related concerns are handled. Beyond the issue of mitigation of damage, the community as a whole ought to benefit in some measure from the overall enterprise. The fortunate leasers of mineral rights should not be the only ones to benefit. That sort of inequity is a sure recipe for neighbor turning on neighbor, particularly when the have-not neighbor incurs some direct negative consequences of the activity.

An interesting thought on revenue generation for broader distribution is the adjustment of valuation of the properties generating revenue from the production. That property is now arguably more valuable on resale than before. (The only argument against is that some will consider production equipment on the property as a negative.) The portion of the increased tax from this aspect could go into a special pot for the purposes discussed above.

In general, taxation related to shale gas production will be the subject of considerable debate. The producing camp will argue that taxation will diminish activity. But anecdotal evidence to date does not support this belief. West Virginia has been much more aggressive in this regard than Pennsylvania, and the drilling activity comparisons are in West Virginia's favor. These single-variable analyses are fraught. In this case, the higher activity in West Virginia may be in spite of the taxation, and driven by the higher quality of the fluids.

Separate from these arguments is the inherent societal benefit from affordable energy. In that sense, the very fact of cheap natural gas represents a benefit for all.

Technology Forks in the Road

Technology choice can often have a direct effect on the local populace. These forks in the technology road fall into two broad categories: benefitting the local environment and aiding the local economy. The first one is an easy choice if other things are about equal. An example of that is in fracturing operations associated with oil or gas production. As the industry became more skilled at drilling horizontally, the increasing reach of a given well allowed a new technology, known as pad drilling. This involves drilling and producing from up to 25 wells from a single location, known as a pad. The number of roads needed drops as does the areal extent of the effects of traffic. Also, this aggregation of wells allows for better supervision and oversight to minimize mistakes. Pad technology was developed in Colorado for the express purpose of minimizing road footprint. It now is even more important in farming communities such as in Pennsylvania.

Biofuels could face similar forks. The conventional approach would be to transport the biomass or crop great distances to giant chemical processing plants. The energy densities of these materials are inherently low. Consequently, the costs and logistics of transportation are high. Technologies are being developed to build more biomass processing plants close to farmland, to bring the mountain to Mohammad, as it were. These must be specialized to not incur the penalties of reduced scale. A few such are in early-stage development. This will not only reduce road transport, but also it would create local jobs, which in many instances are high-paying ones.

Distributed power is another example. Small 50- to 100-megawatt plants using biomass, wind or mini-nuclear, to name a few, could supply localities. In the limit they could eliminate the need for costly and unsightly transmission lines. At short distances, direct current (DC) would be a viable and preferred option to alternating current (AC). Edison would have smiled. For those not familiar with this arcane bit of technological history, this country faced a critical decision in the late years of the nineteenth century. The AC/DC choice was on the table for long-distance power transmission. Pitted were Westinghouse Electric and Thomas Edison. Edison lost that one.

Not farfetched is the notion of developing technologies specifically fit for the purpose of benefitting host communities and nations. An example relevant to the latter would be the deliberate design forks taken to allow for manufacture in that country. One could accomplish this by selecting materials locally available. A design assembly could be modified to use low-grade steel as opposed to high-alloy steel that would have to be imported. Assemblies could be simplified to allow local assembly. In chapter 12 I discuss the distributed processing option for monetizing the ethane found in wet shale gas. That particular option is somewhat hamstrung by the immediate absence of a business model to execute. But these things are traversable, and there is little doubt that local jobs will be created.

In summation, the societal benefit component of energy alternatives need not be an afterthought. Many elements can be brought to bear with no adverse consequences to the economics of the enterprise. Also, the lasting value of being a good citizen cannot be underestimated. It's simply good business.

Flex-Fuel Fairy Tale

"Dream on"

—From "Dream On" by Aerosmith (written by Steven Tyler)

The Utopian State, known the world over as the US, was in the throes of a dilemma. Much maligned for not doing enough to limit carbon dioxide emissions, it developed a plan that seemingly in one fell swoop tackled global warming associated with automobile emissions while at the same time reducing import of oil from nations, some of whom were deemed unfriendly, at least in the rhetoric of elections.

This solution was known as the 20/10 plan. The goal, to replace 20 percent of gasoline with ethanol in 10 years, was seen as visionary, if for no other reason as that 20/10 was about as good as one got with vision. However, even before vast quantities of alcohol had been consumed, a hangover of major proportions was in the making. Therein lies the tale.

The Utopian State, as befitted its name, was inclined to believe that the public would recognize a really good thing when they saw it. They especially believed in the maxim "If you build it, they will come," because said maxim was irresistibly derived from the powerful combination of Kevin Costner, the National Sport, and mysticism.

So they built it, a complex web of subsidies to farmers, automobile companies, and refiners, and tariffs on imported ethanol, all designed to produce domestic ethanol to blend with gasoline, and vehicles that would run on the stuff. In a nod to perceived consumer preferences, they incentivized the auto companies to make flex-fuel cars, capable of using regular gasoline and also E85, a blend with 85 percent ethanol.

They even created demand for these cars by ordering their agencies to use them and mandating the use of the new fuel. Waivers to the mandate were given generously, no doubt in the Utopian belief that said waivers would not be sought if not merited. It seems that some of these agencies are seeing a net increase in gasoline usage (Kindy & Keating, 2008), a result contributing in no small measure to the aforementioned hangover.

At the core of Utopian belief is that folks will "do the right thing." So, purchasers of flex-fuel vehicles were expected to purchase E85, even from filling stations some distance away, ignoring the fuel consumption getting there and back. Then word filtered through that E85 delivered 28 percent fewer miles per gallon. In short, it was more expensive to use and harder to find. They started filling up with regular gasoline because the flex-fuel vehicle allowed that, and filling stations noted the drop in volume and stopped stocking E85.

This nightmare scenario was interrupted by the seemingly sudden realization that natural gas could be produced very cheaply from shale, a rock previously deemed too hard to produce from. US industry knew how to routinely convert natural gas to methanol. This type of alcohol could also be blended with gasoline to make an E85 analog dubbed M85. It was even worse on gas mileage than E85. But methanol from shale gas was so much cheaper than gasoline that the cost per mile driven was less. One had to refuel more often, but at least the public was given the choice. Choice was good but not compelling enough for filling stations to change their design.

The turning point came when Prof. Wunderbahr from a prestigious Eastern university invented a small engine that led to a car running on M85 that delivered both fuel economy and the muscle of a larger engine. The design took advantage of the high octane number of methanol (117 versus 87 for regular gasoline), which allowed effectively high compression ratios, which in turn improved the efficiency of combustion. The result was elimination of the gas mileage penalty from using methanol, increased power for an engine of given size, and retention of the improved emissions associated with methanol usage. And all of this was achieved with a fuel that was consistently less than half the price of gasoline.

Auto makers vied with each other to retool and produce these cars without any federal incentive because the public actually wanted them. Fuel distributors rushed to install M85 pumps and realized that this was simply achieved by eliminating one grade of fuel. They came to the realization that all vehicles on the road today specify either 87 or 91 octane. A third grade (89) was not needed, and the third pump was now available with modification to dispense M85. The US government, not wanting to be left out of this, set policies to further these steps.

Shale gas development technology improved to where low-cost natural gas was assured for decades. This certainty with respect to methanol price staying low allowed massive investment in vehicles and methanol production and delivery infrastructure. All was well again.

And then they elected a president who resolved never again to set policy that was not market-based. The country united behind him on this, and it was never quite the same again. The country was henceforth known as the United States.

PART V

Next
Steps

Research Directions

"A new day will dawn for those who stand long"
—From "Stairway to Heaven" by Led Zeppelin (written by Robert Plant)

The intrusion of a new reality always brings with it the opportunity for significant research. The sheer newness has associated uncertainties and in many instances also disbelief. This latter was certainly the case with Stanley Prusiner's discovery of prions, at the University of California, San Francisco. The idea of infectious proteins was met with a reaction close to derision from colleagues and bordering on vilification by the scientific press. The prominent journal *Cell* rejected his first paper on the subject. Prusiner and colleagues eventually got it published in the *Proceedings of the National Academy of Sciences*, in principle an even more prestigious journal. But a directed review by a member of the National Academy of Sciences is one mechanism (the member, not an independent editor, picks the reviewers)—when this happens it is not classic peer review.

But publishing in the journal still brings prestige because the members, the *crème de la crème* of scientific society, are trusted to adhere to scientific rigor in the directed review process. This is sometimes subject to falling prey to impassioned beliefs. Then again, as in this case, groundbreaking science may see the light of day and not be buried by reviewer orthodoxy.

The idea of producing hydrocarbons from a source rock cannot possibly be considered to be in the same league as the discovery of prions. But it shares the attribute that much research activity will be spawned. Unsurprisingly, the major emphasis will be on improving economics of production and minimizing the environmental impact. Beyond this, the sheer abundance and sustained low cost of this resource will drive innovation in the chemical processing and transportation areas. Natural gas fluids will now be an attractive feedstock.

The Prion Story

At the time it was considered impossible that a protein could replicate and that it could transmit disease. Interestingly, the Nobel Prize was awarded to Stanley Prusiner in 1997 before causality was properly established. Criticism for that decision was immense; one science writer likened it to Henry Kissinger's being awarded the Nobel Peace Prize, which is a low blow of major proportions. Basically, people took sides.

Now there is no doubt that bovine spongiform encephalopathy (mad cow disease), Creutzfeldt–Jakob disease, and scrapie (in sheep) are all transmitted by prions. And the case for the dreaded species-to-species transfer has been made.

Even before the solution was seen as groundbreaking, the problem of transmissible spongiform encephalopathy was seen as important enough to merit another Nobel Prize in Physiology or Medicine for its elucidation. In 1976 Carleton Gajdusek was awarded the Nobel Prize. He was studying the disease kuru and hypothesized that the disease was transmitted due to a custom of a particular Papua and New Guinea tribe. This essentially involved honoring the death of a family member by eating a portion of that person's brain. The disease existed only among the 65,000 tribal villagers in one set of valleys. Gajdusek also popularized the notion of a slow-acting virus being the cause, and this, too, was considered controversial because no such species had been isolated.

In an interesting parallel to the award to Prusiner, the Nobel was given before definite proof of the hypothesis. But his award spurred an intense search for slow acting viruses to cure slow developing illnesses such as cancer and Alzheimer's. The investigators hit a dead end not long after the Nobel award. This is likely the only instance in which a Nobel was awarded first for the wrong explanation for a disease mechanism and then later for the right one.

Improving the Productivity of Wells

This important problem is already being heavily researched by industry. Below are just some of my personal favorites regarding directions to be taken.

The steep production decline rate in the first two years or so has been the subject of much conjecture by opponents of shale gas. As noted earlier, this is not a big factor if the overall economics work. But understanding and ameliorating this phenomenon could produce positive returns. One approach already being investigated is that of a proppant that achieves greater penetration than the conventional. This is responsive to the hypothesis that insufficient proppant coverage in the fractures is allowing earth stresses to close the cracks. This progressively chokes off the flow. An extravagant approach would be to throw the conventional book out. No proppants. Figure out some other way to hold the rock faces apart and allow fluid flow.

I discussed refracturing in chapter 10 and mentioned that there is ongoing research in this area. I have a personal belief that harnessing this may be more fruitful than improving the longer term flow characteristics of fractures. This is because of the inherently tight nature of the rock. No matter how you improve the conductivity of the fractures, the fact that the rock not immediately proximal is unlikely to drain is something to be dealt with. Maybe a combination of technologies is the answer. It will come down to cost-effectiveness. Wet gas, and by extension shale oil, can tolerate a lot of cost if the benefit is there.

Gas to Liquids

Predictably low gas prices combined with the desire to reduce oil imports ought to drive GTL activity. The conventional Fischer-Tropsch process needs improved economics to be able to be competitive with episodic drops in oil price. The North American outlook, and stranded gas in Alaska in particular, should be sufficient impetus for the needed research, most likely in the catalysis aspect. An interesting area for development is that of small-footprint conversion of methane to liquids. This would enable distributed conversion, as mentioned in earlier chapters. But the readily realizable product is a gasoline substitute, diesel being more complicated. The preponderance of gasoline usage in North America for passenger vehicles takes the sting out of this limitation.

The small-footprint conversion of ethane to ethylene is also a target. This is by far simpler in processing terms than the methane conversion mentioned above because ethane already has a carbon-to-carbon bond, and ethylene is simply ethane with two hydrogen atoms removed. Here, too, the principal challenge is innovation in catalysts, although advances will also be needed in the effective separation of constituents, particularly in a portable configuration.

A more mundane but important area is the removal of minor constituents prior to any of this processing. This primarily involves removal of sulfurous gases and carbon dioxide. Alaskan gas sometimes contains up to 12 percent carbon dioxide, although Lower 48 shale gas is advantaged in this regard (with some exceptions). The research again will likely center on small-footprint and minimally energy-intensive methods.

Environmental Issues

These fall broadly in the categories of matters related to water usage and disposal; measurement and control of air emissions; and measures to ensure best practices in all of these.

Water withdrawals: To minimize fresh water usage, saline waters of convenience ought to be employed. The research to tolerate salinity in fracturing fluids has been done to a point. An interesting objective would be to pilot the use of reverse-osmosis reject brine as the base fluid. A quick explanation: reverse osmosis is the current workhorse desalination method. A starting concentration of sea water would end with fresh water and heavy brine. The latter needs to be discarded, and this is problematic in some instances.

The more likely saline water of convenience is from saline aquifers, as discussed earlier. The research needed is in characterizing those proximal to shale reservoirs. At a minimum the constituents evaluated would be bacteria and some of the different salts. A broad effort in each target state is called for. Once characterized and made available to the public, the steps of cleaning and use ought to be straightforward.

Flowback water disposal: In areas with no deep disposal well capability, the problem is significant. All of the Marcellus and Utica for sure have no easy disposal capability, and this has been the cause of serious angst in Pennsylvania. The research will likely focus on the most cost-effective means to reuse the flowback water for fracturing. A lot of work has gone into this aspect, so in the main it will probably be just development activity. Having said that, a better mouse trap could speed the reuse. This probably again will be in small-footprint or portable systems. Large water treatment facilities that aggregate the flowback water from multiple wells will not be the favored route unless there are specific, compelling reasons. Movement of this water in trucks is subject to accidental spills, and the farm roads would welcome less traffic.

Fugitive emissions of methane and, separately, volatile organic compounds have acquired some currency in discussion. While many of these are amenable to simple good practices, there could be scope for better measurement and reporting systems. In April 2012, the US EPA issued a comprehensive set of regulations that primarily address release into the atmosphere. These ought to be actionable by industry without much research. However, the most celebrated of the fugitive methane issues are those of aquifer contamination. Release of detailed plans for testing water wells before drilling (baseline) and at regular intervals during and after drilling is called for. Systems need to be in place to advise homeowners unfamiliar with their rights.

Capturing and utilizing the gas associated with flowback water during the early days of the wells could be a target. To do so economically could be quite challenging. The vented gas onboard LNG ships is utilized effectively because engines exist that can usefully employ the gas. In shale gas operations, switching operating equipment fuel to natural gas would be one target. But that may not consume enough. A high-value target would be the small-footprint conversion of the methane to something useful. Since it will be difficult to make this economic, inducements could be offered to further this sort of endeavor.

Social Science Research

Given the national importance of shale gas, research must be conducted on customer perceptions and understanding of the pluses and minuses. Beyond this, communication campaigns ought to be devised to close the gaps in understanding. This will dampen the confusion caused today by stridency from both sides of the debate.

A valid area of research is systematic economic development methodology as relates to shale gas. In many instances shale gas drilling is in areas new to such industry. Many of the players are small companies that may need guidance in best practices with regard to considering societal benefit in the sustainability equation.

Finally, best practices on casing and cementing need to be detailed, as do methods to track and ensure these practices. Again, this is not the province of research, just something that needs to be done.

Policy Directions

"Come senators, congressmen, please heed the call"

—From "The Times They Are a-Changin'" by Bob Dylan

Shale gas has burst into our energy consciousness with such rapidity as not to allow for a great deal of preparation in the form of informed policymaking. The economic prosperity message has been blunted by the environmental risk cacophony. Both aspects are in need of decided action by all concerned parties in order for the nation to enjoy the economic windfall while still managing the associated risks.

While rulemaking is important, equally critical is voluntary joint industry action. When the industry embarked upon challenging deepwater exploitation, it formed the consortium DeepStar to jointly address the critical hurdles. The environmental issues facing the shale gas industry do not even begin to approach the difficulty of the deep water tasks. Not to minimize the importance, but with a couple of exceptions, the bulk of the actions require the broad scale execution of best practices. This task can and must be taken on, in part because unconventional resources have, until recently, been the sole province of smaller companies with limited resources.

When the Research Partnership to Secure Energy for America (RPSEA) was formed in 2006 with congressional line-item funding, the two targets were ultra deep water and unconventional resources. But even then, when the shale gale was a mere breeze, Congress insisted on a "small operator" provision, meaning that at least some projects were required to have a clear line of sight to benefit the small producer. Today the nation would be well served if RPSEA's focus were to be exclusively tight oil and gas, the latter including shale gas. Strictly from a national imperative, all the evidence now points to tight gas on land being a more important resource than ultra deepwater gas. The definition of tight gas includes shale gas and gas found in other low-permeability rock such as sandstones and carbonates. Also, with very few exceptions, ultra deepwater gas is the domain of the big players. They can afford to finance research on their own, or through consortia, such as the recent consortium to devise spill containment technology, in response to the Deepwater Horizon disaster.

The use of public funds in implicit support of any major profit-making industry is fraught. Therefore, RPSEA funding of ultra deepwater efforts is questionable. Risk reduction in shale gas is different. It ought to be considered a national priority because of the economic gain that would otherwise be at risk. Also, the primary beneficiaries, aside from communities in which production happens, will be smaller producers. It is true that this landscape is continually changing, with bigger players buying into the asset base. But the charge that support of shale gas is support of Big Oil has little merit in the main. The list of producers in Pennsylvania today is a Who's Who of companies that the public never heard of. Ironically, the entrée of bigger players is likely to improve operational diligence. Some may not like the size and profitability of ExxonMobil, but for operational discipline they have no equal.

The ideas expressed below are intended to stimulate discussion and not be overly prescriptive. The most important areas are those that ensure environmentally secure operations. Within each category, though, my ideas are addressed in no particular order.

Environmental Issues

Preventing well water contamination: The most important measure is to ensure the collection of baseline data on a specified set of chemicals including methane in all wells in a predefined proximity to the drilling operations, likely 2,500 feet. The appropriate authority (federal, state, or county) should specify the nature and frequency of the testing and qualify the testing entities. The cost ought to be borne directly or indirectly by the producer. The purpose is to definitively assess whether fugitive methane from the production operations is entering well water. In the event such a leak is detected, regulations ought to be in place for remedial action to fix the leak.

Well operators should be required to place cement sealing over every zone of potential gas production above the reservoir. Failure to do so incurs the risk of a source for leakage. This cement job and the normal casing and cementing operations ought to be verifiably competent, as determined by appropriately specified testing. Industry best practice in this area should be documented and made available to all producers. Regulatory oversight for compliance may be needed. This can be facilitated by the use of sophisticated sensors and means to transmit via satellite the data to operational supervisors as well as regulators in central urban locations, thus minimizing oversight cost.

Handling flowback water: Regulations must be in place for safe handling of flowback water. The most preferred options are to treat and reuse, and disposal in UIC Class II wells. Consideration could be given to making mandatory only these two options, with variances granted only on a case-by-case basis. Disposal in UIC Class II wells is less expensive than treating and reusing flowback water when available. However, further diligence on such wells relative to procedures to minimize the risk of interaction with active faults is needed. Failure to do so can lead to minor earthquakes. Reusing flowback water will be facilitated by a practice of using salty water as the base fluid for fracturing. Even so, service providers will need to gear up to provide the treatment service, so benign alternatives may have to be found for each area in the interim.

Minimizing fresh water usage: Industry today can feasibly utilize salty water for fracturing. Consideration should be given to regulations requiring this as the default, with explainable variances only in cases where salty water of convenience is absent or temporarily unavailable.

Disclosure of fracturing chemicals: Full disclosure of all chemicals in fracturing fluids should be mandatory. The proprietary product exception must not be permitted to be used as a shield or an artifice. The nonprofit website FracFocus ought to be considered as a simple vehicle for disclosure and public access.

Minimizing fugitive gas emissions: Gas produced prior to the hooking up of a pipeline to the production site must be handled in the most environmentally secure fashion. In chapter 23, I recommend innovation in small-footprint, preferably mobile, means to capture the gas and convert it to a transportable liquid. For economics purposes this gas could be considered to be zero cost.

Absent this, or in the interim until available, consideration ought to be given to flaring the gas or most preferably using it for heat or another useful purpose. Direct release is the most environmentally deleterious option.

Pad drilling: This is a mode of production wherein multiple wells are drilled from a single location. This facilitates all of the recommendations made above and has a positive environmental footprint with respect to roads and traffic. It is a net benefit for the producer as well because a few experienced operators can now supervise several wells. Best practices must be shared to nudge pad activities to commence as early in the development as possible.

Economic Issues

Ethane monetization: State and local governments should consider inducements for local processing of the ethane, be they regional crackers or pad-level reactive options (see chapter 12). The compelling economics may be sufficient, but the governments should also take full benefit for local workforce training and the like. Consideration ought also to be given to encouraging research and development centers in the Marcellus/Utica region targeting research related to wet gas. Proximal production would allow easy access to field testing sites.

Export issues: Abetted by a warm winter, abundant shale gas has caused the price in early 2012 to be the lowest in a decade. Even if this is ignored as aberrant, there is little doubt that a lot of cheap gas will become the norm. The temptation to export the gas is high, and this is currently being debated. The box below gives some views on this matter.

LNG Export Is Not in the National Interest

We must not export natural gas in any form in favor of producing and exporting a higher value product. The single most valuable such high-volume product is ammonia-based fertilizer. (Carbon black would be higher value but is a smaller market.) Until recently the US imported half the fertilizer consumed. This is because variable and high prices in the early part of the century caused many manufacturers to relocate abroad to areas of cheap gas such as the Middle East. Now with the prospect of cheap and stable shale gas, many of these are returning. No doubt the chemical industry is skittish about LNG export concepts because exports could vitiate the business assumptions of low cost, were the prices to rise due to massive export of gas. One permit recently sought by Cheniere Energy is unlikely to have a big effect, but many such could.

Aside from the pricing issue, another reason to export product rather than gas is simple economics. Take the example of anhydrous ammonia, the basic building block for nitrogen fertilizer manufacture. About 33.3 mcf gas converts to 1 ton of anhydrous ammonia. The gas value, using $4 per mcf, is $134. The value of the anhydrous ammonia is in the vicinity of $800. Also, domestic labor was used to get it to that state. The landed price in Europe of gas as LNG will be about $7.50 with $4 gas. That near-doubling of value added does not contribute much to the domestic economy. Even the ship was probably made in Korea.

The US will be one of the lowest cost producers of ethane-based ethylene and derivative polymer in the world. This raises the high likelihood of the US being a net exporter of these chemicals. If gas prices stay low, wet gas will be produced almost exclusively. This could cause an ethane glut and lead to exports of ethylene derivatives. This ought to be permitted and encouraged.

States proximal to wet gas production but without any of their own, such as North Carolina, ought to consider encouraging chemical production. Wilmington, North Carolina, already has a chemical industry and being a port could be an export site. Prosperity from cheap and abundant gas does not have to be restricted to the producing states.

Oil substitution: Three principal avenues present themselves: electric vehicles, natural gas replacement of gasoline and diesel, and conversion of natural gas to liquids. The potential policy drivers for this could include:

- Decision by Alaska to kill the gas pipeline to the Lower 48 and to encourage the conversion of vast quantities of cheap stranded gas to liquids. The owners of the gas, principally major oil companies with a thorough understanding of GTL technologies, ought to be motivated to do this with no inducement. Absent this the TAPS, the pipeline bringing oil down from the North Slope, is at risk of closure, as discussed in chapter 13.

- Shale gas–producing states should consider requiring a displacement of diesel with DME or methanol for the pressure pumps used in fracturing and cementing, and natural gas or methanol for the vehicles. There will be issues of access to refueling, capital cost of retrofits and the like, and at least the refueling could be addressed by the states. This could be a bit tricky because the fuel of choice for trucks would be LNG, not CNG.

- All metropolitan areas ought to consider emulating Delhi and a host of other Indian cities where all public transport switched to CNG. The World Bank evaluated the health benefits as very significant, as noted in chapter 15. But from the standpoint of reducing imported oil, replacing gasoline has a bigger bang and methanol may be the route, as discussed in chapter 16.

- Dimethyl ether (DME) can substitute for diesel up to at least 20 percent with no engine modification, emits zero particulates, and has a very high cetane rating. Cheap natural gas equates to cheap DME. States should consider DME additive to diesel as an alternative to a wholesale switch to natural gas or as an early step toward that goal. New plants for DME production will be required. Since DME is a single processing step beyond methanol production, methanol and DME strategies can be pursued simultaneously.

National Security

The military is a gigantic user of transport fuel. Relative to fuel switching, on the one hand mission criticality would demand the most proven and reliable fuel. On the other hand, the delivery of such a fuel to the front lines has huge elements of cost and loss of life. The military ought to conduct an analysis of options, some considerations of which are listed below:

- Base operations are amenable to a complete switch to CNG-fueled vehicles. In fact it may not be far-fetched for military vehicle design to take advantage of methane's 125+ octane rating with a high-compression engine. If the military standardized on such an engine, a civilian version down the road could be feasible. Think Hummer.

- Bases, particularly in foreign countries, could operate small-scale electricity generators using natural gas with stored backup of CNG or even methanol. This could be augmented with solar power in appropriate locations. Nonreliance on an outside grid would have security advantages.

- Aside from the physical risks in delivering fuel and water to the front lines, the cost of security for the convoy likely raises the cost of these commodities to many times the normal. Consequently, the military could consider sponsoring research in the small-footprint distributed production of transport fuel using the raw material of convenience. The baseline cost for such fluid would be the security-inflated cost of convoy delivery. Also, a switch to electricity for vehicles where feasible should be strongly considered. Electricity production in forward locations is more feasible than liquid fuel production.

Finally, abundant shale gas in North America has the effect of more LNG delivery to Europe. This significantly reduces Russia's ability to use gas supplies as a weapon of political will. One could expect increased US influence worldwide as a result. While no specific policy actions spring to mind, lawmakers ought to be aware of the consequences in this regard of significantly holding back shale gas production in the US.

Conclusions

"There will be an answer, let it be"
—From "Let It Be" by The Beatles (written by John Lennon and Paul McCartney)

Shale gas has the potential to materially improve the economic lot of every citizen of the US. There is also the real possibility that, together with distributed cleaner energy such as wind and solar, cheap energy in the form of shale gas could improve the human condition worldwide. Such lofty rewards beg for a concerted effort by all concerned to solve the associated environmental risks. In the opinion of this author, that is an acceptable risk provided we employ a combination of innovative technology and regulatory oversight, and provided the industry will to do the right thing, nudged along by informed local activism in the locations where the industry operates.

The public has a right to know the timeframes involved in the rewards and the risks. While the intent of the book is to allow the readers to form their own opinions, I will attempt a response to the question.

The reduction in the cost to heat homes was already felt in the winter of 2011; one estimate is a $1,000 average reduction per household each year. This trend can be expected to continue.

Ethylene and derivatives and low-cost fertilizer will make their presence felt on the economy in the short term, two to five years. The quicker results will be from resuscitating previously mothballed plants. To the extent that there is a world price for ethylene and derivatives and low-cost fertilizer, export could dampen the direct benefit to farmers and other consumers. Policy measures could ensure the domestic benefit. In any case, the balance of trade would improve if for no other reason than through the reduction of imports.

Methanol displacement of gasoline and diesel is an exciting consequence of shale gas production. The needed legislation to require most cars to accept any combination of conventional fuels and alcohol could pass this year or the next. The compelling economics could drive change in the two to five years. The slowest link in that chain will be adding methanol capacity and establishing fueling infrastructure. The high-compression engine which effectively would more than halve the cost of fuel will take longer—up to 10 years. This timeframe could move up if the public is vociferous in demanding it.

The positive effects on US national security for dimming the aspirations of Russia and Iran would occur in the two- to five-year time frame. A reduction in imported oil would take longer, closer to 10 years for a material effect.

On the environmental risk side, the public will want to know whether measures to manage the risk are working and how soon this will be known. Regulations requiring that flowback water be either reused or disposed of in UIC Class II wells only ought to be in place by next year, if states choose to impose them. Putting in place measures to assure prevention of earthquakes from disposal well activity ought to be a short-term activity. Proper measures ought to eliminate this phenomenon. Technology for reuse of flowback water already exists, but more will be needed to broaden its use and reduce costs. Re-use is the more desirable of the two options. Full-scale implementation ought to be possible in two to five years.

Since programmed releases to the surface will not be an option, any accidental spills during the two allowable handling methods would be observable events. With the industry trending toward the use of greener constituents in fracturing fluid, the only substantive issue with accidental spills should be the saltiness. Environmental degradation just from high salt would not be as long lasting as it might have been from organic compounds such as diesel, which would be forbidden from use in that application.

The use of salty water in place of fresh for fracturing is technically feasible today. Implementation ought also to be in the short term provided there is the will to do it. State legislation requiring the use of salty water would be desirable, and community activism would be helpful in persuading legislators.

Preventing well water contamination ought to be straightforward if my suggestions, based in part on the recommendations of the Pennsylvania governor's Marcellus Shale Advisory Commission, are followed. A key aspect is that industry must make best practices on well construction and monitoring available to the smaller producers. Baseline testing of proximal water wells at industry expense prior to drilling activity, followed by routine testing after, will provide definitive proof of efficacy of well construction. There would no room for uncertainty. In the event of an observed leak, response time for remediation ought to be weeks, not months. From an operational standpoint, sound well construction preventing fluid leakage is just a matter of following good practice. Consequently, failures to do so ought to carry severe penalties.

Methane releases during early stages of gas production are believed to occur in some instances. In each case, the gas is automatically collected as part of a process for separating it from the flow back water. So the operator knows how much there is and precisely where it is stored. This potent greenhouse gas must not be directly discharged. If no useful purpose is found, it ought to be burned on location (flared). Whether this gas is used or flared, the public ought not to have concerns on this score. If operators are required to report on all dispensation of the gas other than commercial export, the public would realize an even greater measure of comfort without undue burden on the operator. Automatic monitoring with remote reporting ought to be straightforward.

There is little doubt that shale gas is transforming the US energy-based economy. An importer of natural gas, with imports of 10 percent of need in 2010, we are headed to an era of domestic self-sufficiency. The argument has shifted to whether to permit export of liquefied natural gas. The sheer abundance of gas is keeping prices low, and models from two different sources indicate that the prices should remain low to moderate for decades. An important aspect of note is that the prices can be expected to remain stable, without the peaks and valleys experienced in the past. This stability is almost as much an attraction to industrial users as the low pricing is.

Even at the upper end of the modeled pricing range, $8 per MMBTU, gas-based electricity will be cheaper than that from new coal plants, even without a price on carbon. However, if prices stay in the lower end of the range for long, renewable energy sources will be slow to be adopted unless there is policy intervention. Another consequence of sustained prices at the low end will be emphasis on wet gas production, because the majority of the profit will be in the wet component. This will cause a glut in ethane, which, if properly anticipated through cracker capacity addition, could make the US the low-cost producer of ethylene, allowing for lucrative exports.

Low-cost energy is a tide that lifts all boats of economic growth. Shale gas is a powerful such tide. It has burst upon us so unexpectedly that we have become rattled by the flotsam it carried with it. I conclude that the flotsam is manageable, allowing us to enjoy the benefits of the tide.

Bibliography

These represent citations made in the essays plus other suggested reading. Inevitably many of the sources are news stories and blogs. In using the information from each of these, I have considered the source for possible slant or bias. But readers ought to draw their own conclusions in this regard. Many of the assertions made in the essays are drawn from my experience. In these cases no citations are made.

Akbar, S., Lvovsky, K., Kojima, M., & World Bank South Asia Region. (2005, June 1). For a breath of fresh air: Ten years of progress and challenges in urban air quality management in India, 1993–2002 (English). Working paper 35047. New Delhi: World Bank Environment and Social Development Unit, South Asia Region, Retrieved March 19, 2012, from http://documents. worldbank.org/curated/en/2005/06/6561544/breath-fresh-air-ten-years-progress-challenges-urban-air-quality-management-india-1993-2002

Alyeska Pipeline Service Co. (2010). Trans Alaska Pipeline System quick reference guide. Retrieved May 29, 2012, from http://www.alyeska-pipe.com/ Pipelinefacts/TAPS%20Quick%20Facts.pdf

American Chemistry Council. (2011). *Shale gas and new petrochemicals investment: Benefits for the economy, jobs, and US manufacturing.* Washington, DC: American Chemical Council. Retrieved March 14, 2012, from http://www.americanchemistry.com/ACC-Shale-Report

Barcella, M. L., Gross, S., & Rajan, S. (2011). *Mismeasuring methane: Estimating greenhouse gas emissions from upstream natural gas development.* IHS CERA. Cambridge Energy Research Associates. Retrieved from http://heartland. org/sites/default/files/Mismeasuring%20Methane.pdf

Benchaar, C., Pomar, C., & Chiquette, J. (2001). Evaluation of dietary strategies to reduce methane production in ruminants: A modelling approach. *Canadian Journal of Animal Science, 81*, 563–574. doi: 10.4141/A00-119

Blauch, M. E., Myers, R. R., Moore, T. R., & Houston, N. A. (2009, September). *Marcellus Shale post-frac flowback waters—where is all the salt coming from and what are the implications?* Presented at the 2009 Society of Petroleum Engineers Regional Meeting, Charleston, WV.

Blauch, M., Houston, N., Seyman, M., & Reese, R. (2009, September). Technique reuses frac water in shale. *The American Oil & Gas Reporter.* Retrieved March 13, 2012, from www.swsi.com/pdf/AOGR_Eprint_0909. pdf

Braziel, R. (2011, March). Infrastructure projects connect Marcellus Shale to ethane, NGL markets. Retrieved from *American Oil and Gas Reporter* website: http://www.aogr.com/index.php/magazine/cover-story/ infrastructure-projects-connect-marcellus-shale-to-ethane-ngl-markets Bromberg, L., & Cohn, D. R. (2008). *Effective octane and efficiency advantages of direct injection alcohol engines.* MIT Laboratory for Energy and the Environment, Report LFEE 2008-01 RP.

Brune, M. (2012, February 2). The Sierra Club and natural gas [Web log post]. Retrieved May 14, 2012, from http://sierraclub.typepad.com/ michaelbrune/2012/02/the-sierra-club-and-natural-gas.html

Brusstar, M., Stuhldreher, M., Swain, D., & Pidgeon, W. (2002). *High efficiency and low emissions from a port-injected engine with neat alcohol fuels.* SAE Technical Paper 2002-01-2743. Washington, DC: US Environmental Protection Agency.

Cathles, L. M., Brown, L., Hunter, A., & Taam, M. (2012, February 29). Press release: Response to Howarth et al.'s reply. *Cornell Geological Sciences.* Retrieved March 14, 2012, from http://www.geo.cornell.edu/ eas/PeoplePlaces/Faculty/cathles/Natural%20Gas/Response%20to%20 Howarth's%20Reply%20Distributed%20Feb%2030,%202012.pdf

Chelme-Ayala, P., El-Din, M. G., Smith, R., Code, K. R., & Leonard, J. (2011). Advanced treatment of liquid swine manure using physico-chemical treatment. *Journal of Hazardous Materials, 186*(2-3), 1632–1638.

Chesapeake's new Utica Shale wells producing "very strong." (2011, September 29). *Marcellus Drilling News.* Retrieved March 19, 2012, from http://marcellusdrilling.com/2011/09/chesapeakes-new-utica-shale-wells- producing-very-strong/

Cohn, D. R. (2012, March). Use of methanol in flex fuel heavy duty trucks. Presentation at Methanol Policy Forum, Washington, DC.

Despite methane emissions upstream, natural gas is cleaner than coal on a life-cycle basis. (2011, August 26). Retrieved March 15, 2012, from Worldwatch Institute website: http://www.worldwatch.org/despite-methane-emissions-upstream-natural-gas-cleaner-coal-life-cycle-basis

Dozier, G., Elbel, J., Fielder, E., Hoover, R., Lemp, S., Reeves, S., Siebrits, E., Wisler, D., & Wolhart, S. (2003). Refracturing works. *Oilfield Review, 15*(3), 38–53.

The economics of LNG. (n.d.). Retrieved from shareholdersunite.com website: http://shareholdersunite.com/the-ioc-files-useful-background-material/the-economics-of-lng/

Economides, M. (2005). The economics of gas to liquids compared to liquefied natural gas. *World Energy, 8*(1), 136–140. Retrieved March 14, 2012, from http://www.worldenergysource.com/articles/pdf/economides_WE_v8n1.pdf

Europeans shiver as Russia cuts gas shipments. (2009, January 7). Retrieved from http://www.msnbc.msn.com/id/28515983/ns/world_news-europe/t/europeans-shiver-russia-cuts-gas-shipments/

The future of biofuels: The post-alcohol world. (2010, October 28). *The Economist*. Retrieved March 15, 2012, from http://www.economist.com/node/17358802

Go figure: A carbon tax crafted right here at home. (2007, March 9). *Calgary Herald*. Retrieved from http://www.canada.com/calgaryherald/columnists/story.html?id=8c3c9760-7cbe-4fab-b00c-1c77243903b6

Gold, R., & Campoy, A. (2011, December 6). Focus on fracking: Oil's growing thirst for water. *The Wall Street Journal*. Retrieved March 12, 2012, from http://online.wsj.com/article/SB1000142405297020452820457700993022284 7246.html

Harrison, M. (2012, January 17). Revised attachment 3: Gas well completion emissions data, URS Corporation report. [Attachment to memo from A. Farrell & B. Thompson to G. McCarthy, US Environmental Protection Agency]. America's Natural Gas Alliance. Retrieved March 14, 2012, from http://anga.us/media/241555/anga-axpc%20nsps%20memo%20revised.pdf

Heintz, J., & Pollin, R. (2011). *The economic benefits of a green chemical industry in the United States: Renewing manufacturing jobs while protecting health and the environment.* Political Economy Research Institute, University of Massachusetts, Amherst. Retrieved March 14, 2012, from http://www.hhh.umn.edu/centers/stpp/research/pdf/UMassGreen-Chemistry-Report_FINAL.pdf

Howarth, R. W., Santoro, R., & Ingraffea, A. (2011). Methane and the greenhouse-gas footprint of natural gas from shale formation. *Climatic Change, 106*(4), 679–690. doi: 10.1007/s10584-011-0061-5.

International Energy Agency (IEA). (2012). Energy security. Retrieved May 21, 2012 from http://www.iea.org/subjectqueries/keyresult.asp?KEYWORD_ID=4103

Islam and oil. (n.d.). Retrieved May 21, 2012, from AmericanEnergyIndependence.com website: http://www.americanenergyindependence.com/islam.aspx

Jackson, R., & Vengosh, A. (2011, May 10). *Strong evidence that shale drilling is risky.* Retrieved March 15, 2012, from http://articles.philly.com/2011-05-10/news/29528421_1_water-wells-safe-drinking-natural-gas

Johnson, J. (2012, March 19). A long view of the energy market. *Chemical & Engineering News, 90*(12), 34–35.

Kaufman, L. (2011, December 17). Environmentalists get down to earth. *The New York Times.* Retrieved March 19, 2012, from http://www.nytimes.com/2011/12/18/sunday-review/environmentalists-get-down-to-earth.html

Kindy, K., & Keating, D. (2008, November 23). Problems plague US flex-fuel fleet. *Washington Post.* Retrieved March 15, 2012, from http://www.washingtonpost.com/wp-dyn/content/article/2008/11/22/AR2008112200886.html

King, G. (2010). *Thirty years of shale gas fracturing: What have we learned?* Society of Petroleum Engineers conference paper 133456-MS. doi: 10.2118/133456-MS

King, G. (2011, May). A closed loop system using a brine reservoir to replace fresh water as the frac fluid source. In *Proceedings of the Technical Workshops for the Hydraulic Fracturing Study: Water Resources Management* (EPA 600/R-11/048; pp. 76–80). Washington, DC: US Environmental Protection Agency.

King, G. (2012). *Hydraulic fracturing 101: What every representative, environmentalist, regulator, reporter, investor, university researcher, neighbor and engineer should know about estimating frac risk and improving frac performance in unconventional gas and oil wells.* Society of Petroleum Engineers conference paper 152596-MS. doi: 10.2118/152596-MS

Levi, M. (2011, April 15). Some thoughts on the Howarth Shale gas paper [Web log post]. Retrieved March 14, 2012, from Council on Foreign Relations website: http://blogs.cfr.org/levi/2011/04/15/some-thoughts-on-the-howarth-shale-gas-paper/

Luft, G., & Korin, A. (2009). *Turning oil into salt: Energy independence through fuel choice.* Charleston, SC: Booksurge Publishing.

Marron, D. (2009, August 21). The disconnect between oil and natural gas prices [Web log post]. Retrieved May 16, 2012, from http://dmarron. com/2009/08/21/the-disconnect-between-oil-and-natural-gas-prices/ Martin, C., Ferlay, A., Chilliard Y., & Doreau, M. (2007, September). Rumen methanogenesis of dairy cows in response to increasing levels of dietary extruded linseeds. In *2nd International Symposium on Energy and Protein Metabolism and Nutrition,* 9-13 September 2007, Vichy, France, pp. 609-610.

Medlock, K., Jaffe, A., & Hartley, P. (2011). *Shale gas and US national security. Energy forum.* Houston, TX: James A. Baker II Institute for Public Policy, Rice University. Retrieved from http://bakerinstitute.org/publications/EF-pub-DOEShaleGas-07192011.pdf

Molofsky, L., Connor, J., Farhat, S., Wylie, A., Jr., & Wagner, T. (2011, December 5). Methane in Pennsylvania water wells unrelated to Marcellus shale fracturing. *Oil & Gas Journal,* 54–67. Retrieved March 14, 2012, from http://www.cabotog.com/pdfs/MethaneUnrelatedtoFracturing.pdf

Mueller, K., & Yeston, J. (Eds.). (2011, December 2). *Don't panic* [Editors' Choice]. *Science, 334,* 1183. doi:10.1126/science.334.6060.1183-a

National Energy Technology Laboratory (NETL). (2009, April). *Alaska North Slope oil and gas: A promising future or an area in decline?* Addendum report DOE/NETL-2009/1385. US Department of Energy. Retrieved March 12, 2012, from www.netl.doe.gov/technologies/oil-gas/publications/AEO/ANS_Potential.pdf

Nikhanj, M., & Jamal, S. (2010, March 23). No more guessing: Gas rig efficiency and spud to sales. *Natural Gas Supply*. Calgary: Ross Smith Energy Group. Retrieved March 14, 2012, from www.rseg.com/main/ sites/default/rsegfiles/No%20More%20Guessing%20-%20Gas%20Rig%20 Efficiency%20And%20Spud%20To%20Sales.pdf

Norwegian Petroleum Directorate. (2010, July 5). Environmental considerations in the Norwegian petroleum sector. In *Facts 2010*. Retrieved March 15, 2012, from http://www.npd.no/en/Publications/Facts/Facts-2010/ Chapter-9

Norwegian Petroleum Directorate. (2011, July 4). Environmental considerations in the Norwegian petroleum sector. In *Facts 2011*. Retrieved March 19, 2012, from http://www.npd.no/en/Publications/Facts/Facts-2011/ Chapter-9

Nosko, A., & Sevce, P. (2010, September 29). The evolution of energy security in the Slovak Republic. *Journal of Energy Security*. Retrieved March 15, 2012, from http://www.ensec.org/index.php?option=com_ content&view=article&id=262:the-evolution-of-energy-security-in-the- slovak-republic&catid=110:energysecuritycontent&Itemid=366

Nuttall, B. (2008). *Scoping outline for testing enhanced gas recovery (EGR) from the Devonian Black Shales of Kentucky using CO2 injection*. Kentucky Geological Survey. Retrieved March 14, 2012, from http://www.uky.edu/ KGS/kyccs/ppt/EGR_shale_scope.pdf

Oil and natural gas sector: New Source Performance Standards and National Emission Standards for Hazardous Air Pollutants Reviews. US Environmental Protection Agency (EPA). 40 C.F.R. Part 63 (2012). http:// www.epa.gov/airquality/oilandgas/pdfs/20120417finalrule.pdf

Olah, G., Goeppert, A., & Surya Prakash, G. (2009). *Beyond oil and gas: The methanol economy*, 2nd ed. Weinheim, Germany: Wiley-VCH.

Osborn, S. G., Vengosh, A., Warner, N. R., & Jackson, R. B.. (2011). Methane contamination of drinking water accompanying gas-well drilling and hydraulic fracturing. *Proceedings of the National Academy of Sciences, 108*(20): 8172–8176.

Pennsylvania Office of the Lieutenant Governor. (2011). *Marcellus Shale Advisory Commission: Report summary*. Retrieved from Pennsylvania Department of Environmental Protection website: www.marcellus.psu.edu/ resources/PDFs/MSACSummary.pdf

PFC Energy. (2009, July). *Unpacking uncertainty: Investment issues in the petroleum sector.* Retrieved from International Energy Forum website: http://www.ief.org/_resources/files/content/events/unpacking-uncertainty/unpacking-uncertainty.pdf

Plastics prices: Pretty pricey polymer. (2011, February 14). *The Economist.* Retrieved March 15, 2012, from http://www.economist.com/blogs/newsbook/2011/02/plastics_prices

Rao, V. (2011, June). Gas positioned to displace coal plants. *The American Oil & Gas Reporter.* Retrieved March 14, 2012, from http://www.aogr.com/index.php/magazine/technology-feature/gas-positioned-to-displace-coal-plants

Ryder, R. T. (2008). *Assessment of Appalachian Basin oil and gas resources: Utica-Lower Paleozoic total petroleum system.* Open File Report 2008-1287. Reston, VA: US Geological Survey.

Sato, Y., Oikawa, H., & Tsuchiya, T. (2010). *Development of a DME (dimethyl ether) fueled heavy-duty truck.* Retrieved March 14, 2012, from University of California Riverside, Environmental Research Institute website: http://eri.ucr.edu/ISAFXVCD/ISAFXVAF/DDFHDT.pdf

Secretary of Energy Advisory Board, Shale Gas Production Subcommittee. (2011, November 18). *Second ninety day report.* Retrieved March 14, 2012, from http://www.shalegas.energy.gov/resources/111811_final_report.pdf

Seddon, D. (2010). *Petrochemical economics: Technology selection in a carbon constrained world.* London: Imperial College Press.

Sharma, M. (2010). Improved reservoir access through refracture treatments in tight gas sands and gas shales [Presentation]. Retrieved March 14, 2012, from Research Partnership to Secure Energy for America website: http://www.rpsea.org/attachments/contentmanagers/3458/07122-41-Re-fracturing_of_Tight_Gas_Shales%E2%80%93UT_at_Austin.pdf

Simmons, M. R. (2005). *Twilight in the desert: The coming Saudi oil shock and the world economy.* Hoboken, NJ: John Wiley & Sons.

Somers, J., & Schultz, H. (2010). Coal mine ventilation air emissions: Project development planning and mitigation technologies. In S. Hardcastle & D. McKinnon (Eds.), *13th United States/North American Mine Ventilation Symposium.* Sudbury, Ontario: MIRARCO. Retrieved March 14, 2012, from http://www.epa.gov/cmop/docs/vam-planning-mitigation.pdf

Switching from coal to natural gas would do little for global climate, study indicates. (2011, September 8). Retrieved March 15, 2012, from http://www.physorg.com/news/2011-09-coal-natural-gas-global-climate.html

Thaler, R. H., & Sunstein, C. R. (2003). Libertarian paternalism. *American Economic Review, 93*(2), 175–179.

Urbina, I. (2011, June 25). Insiders sound an alarm amid a natural gas rush. *The New York Times.* Retrieved March 14, 2012, from http://www.nytimes.com/2011/06/26/us/26gas.html?_r=1&pagewanted=1&sq=marcellus&st=cse&scp=12

US Energy Information Administration (EIA). (n.d.). Maps: Exploration, resources, reserves, and production. Retrieved from http://www.eia.gov/pub/oil_gas/natural_gas/analysis_publications/maps/maps.htm

US Energy Information Administration (EIA). (2010, January 27). Schematic geology of natural gas resources. Retrieved May 22, 2012, from http://www.eia.gov/oil_gas/natural_gas/special/ngresources/ngresources.html

US Energy Information Administration (EIA). (2011, April 5). *World shale gas resources: An initial assessment of 14 regions outside the United States.* Washington, DC: EIA. Retrieved from www.eia.gov/analysis/studies/worldshalegas/

US Energy Information Administration (EIA). (2011, April 29). *Ethane prices trail other natural gas liquids: Comparative spot price movements in natural gas, crude oil, and natural gas liquids.* Retrieved May 29, 2012, from http://www.eia.gov/todayinenergy/detail.cfm?id=1170

US Energy Information Administration (EIA). (2011, September 30). *Global natural gas prices vary considerably.* Retrieved May 29, 2012, from http://205.254.135.7/todayinenergy/detail.cfm?id=3310

US Energy Information Administration (EIA). (2012, January 23). *AEO2012 Early release overview: Energy productions and imports.* Report Number: DOE/EIA-0383ER(2012). Retrieved May 24, 2012, from http://www.eia.gov/forecasts/aeo/er/early_production.cfm

US Energy Information Administration (EIA). (2012, May 9). Henry Hub Gulf Coast natural gas spot price (dollars/mil. BTUs). Retrieved May 16, 2012, from http://www.eia.gov/dnav/ng/hist/rngwhhdW.htm

US Environmental Protection Agency (EPA), Office of Research and Development. (2011, May). *Proceedings of the technical workshops for the Hydraulic Fracturing Study: Water resources management.* Presented at US EPA Technical Workshop for the Hydraulic Fracturing Study: Fate & Transport, Arlington, Virginia.

US Environmental Protection Agency (EPA). (2011). *Proceedings of the technical workshops for the hydraulic fracturing study: Chemical & analytical methods.* EPA 600/R-11/066. Retrieved from http://water .epa.gov/type/groundwater/uic/class2/hydraulicfracturing/upload /proceedingsofhfchemanalmethodsfinalmay2011.pdf

US Environmental Protection Agency (EPA). (2012). Oil and natural gas sector: New Source Performance Standards and National Emission Standards for Hazardous Air Pollutants Reviews. http://www.epa.gov /airquality/oilandgas/pdfs/20120417finalrule.pdf

US Geological Survey. (2003, October). Depth to saline ground water in the United States (generalized from Feth and others, 1965, Preliminary map of the conterminous United States showing depth to and quality of shallowest ground water containing more than 1,000 parts per million dissolved solids: U.S. Geological Survey Hydrologic Investigations Atlas HA-199). In *Desalination of ground water: Earth science perspectives* (USGS Fact Sheet 075–03). Reston, VA: Office of Ground Water, US Geological Survey. Retrieved from http://pubs.usgs.gov/fs/fs075-03/

US Geological Survey. (n.d.a). Earthquake facts and statistics. Retrieved May 18, 2012, from http://earthquake.usgs.gov/earthquakes/eqarchives/year /eqstats.php?mtcCampaign=10798&mtcEmail=27306046

US Geological Survey. (n.d.b). Magnitude/intensity comparison. Retrieved May 18, 2012, from http://earthquake.usgs.gov/learn/topics/mag_vs_int.php

Utica shale—The natural gas giant below the Marcellus? (n.d.) Retrieved from http://geology.com/articles/utica-shale/Warren, W. J. (2011). System and method for fluid treatment. US Patent application No. 2011/0272155 A1. Washington, DC: US Patent and Trademark Office.

Walters, C. (2006). The origin of petroleum. In C. S. Hsu & P. R. Robinson (Eds.), *Practical Advances in Petroleum Processing.* Springer, pp. 79–101. doi: 10.1007/978-0-387-25789-1_2

Warpinski, N. (2009). Microseismic monitoring: Inside and out. SPE Distinguished Author Series, *JPT*, 80–85.

Wynne, H., Broquin, F. D., & Singh, S. (2010). US utilities: Coal-fired generation is squeezed in the vice of EPA regulation; who wins and who loses? *Bernstein Report*. Retrieved March 14, 2012, from http://207.114.134.6/coal/oh/downloads/bernstein-report.pdf

Zoback, M., Kitasei, S., & Copithorne, B. (2010). Addressing the environmental risks from shale gas development (Briefing paper 1). Retrieved March 14, 2012, from Worldwatch Institute website: http://www.worldwatch.org/files /pdf/Hydraulic%20Fracturing%20Paper.pdf

Song Lyrics

Casey, H. W., & Finch, R. (1976). (Shake, shake, shake) shake your booty [Recorded by KC and the Sunshine Band] [7" single]. Miami, FL: TK Records.

Crew, B., & Gaudio, B. (1967). Can't take my eyes off you [Recorded by Frankie Valli] [7" single]. Philips Records.

Croce, J. (1971). You don't mess around with Jim. On *You don't mess around with Jim* [Album]. ABC Records.

DuBois, T. (1982). She got the gold mine (I got the shaft) [Recorded by Jerry Reed] [7" single]. RCA Records.

Dylan, B. (1962). Blowing in the wind [7" single]. Columbia Records.

Dylan, B. (1964). The times they are a-changin'. On *The times they are a-changin'* [Album]. Columbia Records.

Fogelberg, D. (1981). The leader of the band [7" single]. Full Moon Records.

Jagger, M., & Richards, K. (1965). Satisfaction [Recorded by The Rolling Stones] [7" single]. London Records.

Joel, B. (1979). You may be right [7" single]. Columbia Records.

Knopfler, M. (1985). Why worry [Recorded by Dire Straits]. On *Brothers in arms* [Album]. Vertigo Records and Warner Brothers Records.

Lennon, J. & McCartney, P. (1970). *Let it be* [Recorded by The Beatles]. On *Let it be* [Album]. London: Apple Records.

Lennon, J., & McCartney, P. (1965). We can work it out [Recorded by The Beatles] [7" single]. London: EMI Studios.

Lennon, J., & McCartney, P. (1969). *Get back* [Recorded by The Beatles] [7" single]. London: Apple Records.

McLean, D. (1971). American pie [7" 45 RPM single]. Los Angeles, CA: United Artists Records.

Nettles, J. (2006). Stay [Recorded by Sugarland]. On *Enjoy the ride* [CD]. Nashville: Universal Music Group.

Page, J., & Plant, R. (1971). Stairway to heaven [Recorded by Led Zepplin] [EP single 45 RPM]. Atlantic Records.

Porter, C. (1934). I get a kick out of you [Recorded by Ethel Merman]. In *Anything goes* [Musical]. New York: Harms Inc.

Ragovoy, J. (1963). Time is on my side [Recorded by The Rolling Stones] [7" 45 RPM single]. Verve Records.

Sebastian, J. (1965). Do you believe in magic? [Recorded by The Lovin' Spoonful] [7" single]. Kama Sutra Records.

Simon, P. (1969). Bridge over troubled water [Recorded by Simon and Garfunkel] [7" single]. Columbia Records.

Stevens, C. (1971). Peace train. On *Teaser and the fire cat* [Album]. A&M Records.

Stills, S. (1966). For what it's worth [Recorded by Buffalo Springfield]. [7" single.] Atco Records.

Tyler, S. (1973). Dream on [Recorded by Aerosmith] [7" single]. Columbia Records.

Waters, R. (1979). Another brick in the wall [Recorded by Pink Floyd]. On *The wall* [Album]. Columbia Records, Harvest Records/EMI Records, and Capitol Records.

Webb, J. (1967). Up, up and away [Recorded by The 5th Dimension] [7" 45 RPM single]. Soul City Records.

About the Author

Vikram Rao, PhD, is Executive Director of the Research Triangle Energy Consortium (RTEC), a nonprofit in energy founded by Duke University, North Carolina State University, RTI International, and the University of North Carolina at Chapel Hill. Its mission is to illuminate national energy priorities and by extension those of the world, and to catalyze research to address these priorities.

The Research Triangle Energy Consortium (RTEC), representing Duke University, NC State University, the University of North Carolina at Chapel Hill, and RTI International, was formed to have a major impact on solving the nation's and the world's technical, economical, societal, and public policy problems related to the use of energy.

While in a pro bono advisory capacity to a major nongovernmental organization (NGO), Dr. Rao became very familiar with the environmental issues related to shale gas. This added to his previous knowledge of shale-gas-related technology and operations. As an organization, RTEC was increasingly viewing natural gas as a transitional fuel to reach a future dominated by renewables. Also readily apparent was the fact that the nation was deeply divided regarding the ability to safely produce the needed natural gas domestically. Since rhetoric was often overtaking knowledge, Dr. Rao decided to use his background, augmented by more recent research, to illuminate both sides of the debate with a book designed to be readable by the lay public. This is the result.

Dr. Rao serves on the board of Intelligent Well Controls Ltd. and also advises venture capital firm Energy Ventures AS, and firms BioLargo, Inc., Global Energy Talent Ltd., and Integro Earth Fuels, LLC. He retired as Senior Vice President and Chief Technology Officer of Halliburton Company in 2008 and followed his wife to Chapel Hill, North Carolina, where she is on the faculty of the University of North Carolina. Later that year he took his current position.

Dr. Rao is an engineer by training. He received his B Tech in metallurgy at the Indian Institute of Technology, Madras, India, followed by an MS and PhD in materials science and engineering at Stanford University. He has been married for 38 years to Susan J. Henning, a professor at the University of North Carolina at Chapel Hill. They have three sons. Justin, an economist, works in New York City. Colin, a mechanical engineer, works in Houston, as does his twin brother Mitchell, a computer scientist. The Raos' aging dog, Kalu, stands ready to defend their home by lavishing love on any intruder.

Acknowledgments

George King of Apache has been a willing resource and fact- and logic-checker on the upstream aspects. Raghubir Gupta and David Denton, both of RTI International, did the same on the downstream side. But if any errors have crept in, they are to be attributed solely to me. Sara Lawrence, RTI, and Daniel Raimi, Duke University, contributed data and discourse on societal issues.

Gordon Allen read the book and tailored his pen and ink drawings to the message, astutely handling the suggestions of the engineer author.

I am grateful to the RTI Press production crew for putting up with a rookie author—and an engineer to boot. They allowed themselves to be taken out of their respective comfort zones until convergence was achieved. Considering that they were the pros, this was a significant concession.

Many contributed, but special citation goes to Karen Lauterbach, Brad Walters, and Brian Southwell. Brian was immensely helpful early in trashing my first attempt at an Introduction. Brad somehow found time away from his day job to edit the book in timely fashion. And Karen held us all together. The reader will judge whether it all worked.

Made in the USA
Lexington, KY
24 January 2013